THE NEW VITICULTURE

THE

NEW VITICULTURE

Douglas Meador

Wine Grower

ELLEM PUBLISHING, INC.

www.ellempublishing.com

+1-831-626-3035

ISBN 978-0-6152347-4-8

LCCN 2008911722

CONTENTS

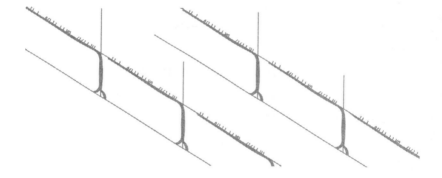

AKNOWLEDGMENTS

Throughout the last thirty-five years three people have had tremendous impact upon my thinking processes and have made considerable contribution to the outcomes.

The first is Terrell West who is probably the finest pure grape grower I have met. His father was a grower and Terrell was raised with grape juice in his veins. Throughout the seventies, as I was endeavoring first to understand and second to elucidate the vinifera grape "system", Terrell was both my "sounding board" and contributor to what eventually evolved.

In the seventies Terrell and I would spend some hours per week in a coffee shop in Greenfield discussing new thoughts. When I would show up with my pads of drawings and equations and proposed solutions to problems Terrell would study them, lean back and say "Nope. Won't work." I would respond "Why"? He would then explain in practical terms from experience and logic. I would go back and rework my thinking. Many of the elements of the resulting ideas on viticulture sprang from his mind however any errors or misdirections within the presented work are purely my responsibility.

Terrell continues to this day to be my "Go To" guy whenever I have technical questions concerning rootstalks, clones, varietals, etc.. He is a veritable encyclopedia of vinifera knowledge.

The second individual who was of immense importance to me in the evolution of the system was Dan McNamara of Quiedan Corporation. I have known Dan since 1965 – first as a buyer for Raychem Corporation then as a buyer for Stanford Research Institute which led him into exotic worlds finding advanced or strange materials requested by those Stanford explorers of the future. Dan took that vast knowledge and put it to work, forming Quiedan in 1976. Through the years Dan would take my sketches, often on bar napkins, of my material needs and make them real. Regularly, he would make suggestions for improvements to my design, suggest materials and possible substitutes. Once decided he would cause their manufacture. The world of metal and its bending or shape forming is very arcane, very technical and is truly mysterious to me. At all times I bowed to his knowledge, simply stating or drawing my needs and he would supply. Of course, as the evolved system became widely accepted others would copy his devices shaving ten cents of strength and selling for a penny less was/is common in the world. Still, to this day, Dan supplies to those who value long-term economy over short.

Colonel Ken Gingras, (USA RET), and his wife Kris have been of immense assistance the last twenty-five years in our wandering of Europe. Ken and Kris have a strong love of wines and the cultures of Europe. Their guidance, companionship and astute comments have been welcome and their non-professional approaches have, at times, attracted my attention to points I may have missed. Kris' fluency in German and Spanish has been particularly helpful. Ken is the one who finally forced me into starting this book telling me one morning as he headed to work that if the outline wasn't finished by the time he returned at noon he would "kick my tush". The outline was finished.

While I could read some German on viticultural subjects my spoken German was worse than horrible and was received with great laughter and sympathy. When we hadn't been in Germany for some time Kris would call begging us to return as she was "running out of Doug stories to entertain guests"!

One example was at a restaurant when I attempted to tell the owner the dinner was superb and I really liked it. Turned out I told him *he* was beautiful and I really liked *him*! It took some rapid talk by Kris to settle the moment! Her fluency was of great assistance in the field.

Of course, no man is an island and I have benefited from a myriad of minds throughout the world. From formal technical treatises of France and Germany to chance thoughts, comments or observations made by practitioners, all have contributed to the evolved theory. I am an equal-opportunity thief of ideas – an applicable idea is just that regardless of its source. I have been so fortunate to be allowed access, in their productive years, to superb minds such as Dick Graff, David Bruce, Ken Burnap, Andre Tschelitsheff, Bob Mondavi, Angelo Papagni and the list goes on and on with too many more to mention. I am grateful to them all, as we all should be, for building the foundation of the American fine wine industry.

I would particularly like to thank with profound gratitude Tina Bayless who, with her internal Rosetta Stone, was able to change my handwriting scribbles into words that most English speakers will recognize. Being computer illiterate, I relied upon her skills to render this text into usable form. Thanks, Tina.

Jennifer Rudolph converted my amateur drawings into professional renditions. Thanks Jennifer and I hope you enjoy the wine.

x

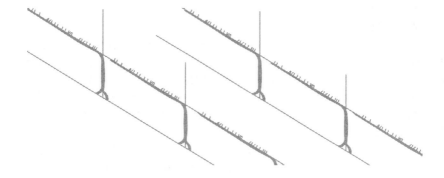

Antonio: What's past is prologue.
Shakespeare: THE TEMPEST

PROLOGUE

On September 15, 2006 LuAnn and I sold The Ventana Vineyard and Winery to a group local to Monterey. They already have vineyards in the area around The Ventana. It was time to "pass the baton" of stewardship of the most award-winning single vineyard property in America. My thirty-four year run as, essentially, a research vineyard has been long and gratifying. Extraordinary levels of productivity can be accomplished by finding the vocation that evokes and conjures up emotional energy and intellectual curiosity and challenge. I was very, very fortunate in that respect and have been in a love affair with the wine world all that time. LuAnn and I are blessed with friends throughout the world and we cherish each and every one. We have been immersed in so many different cultures and each has affected us deeply. Perhaps now there will be more time to visit again.

I did not start out on this adventure with any idea that my work would change viticulture as it then existed, nor did I remotely dream that I would become so dis-enchanted by the existing academia and its methodology that I would eventually have the temerity to recommend a re-organization of the academic system. It is amazing to me that this poor farm boy's search for understanding would lead through languages, countries, history, chemistry, plant physiology, finance, marketing, winemaking, forestry, cooperage, bacteria, yeasts, engineering, etc., while also being given the opportunity to make friends and acquaintances throughout the world.

In this pursuit when faced with diversity I sought unity. I really could not tolerate the segmentation of knowledge that is characteristic of most of academic work. After some time spent familiarizing myself with the subject I felt progress in fine winegrowing was severely hindered because the research community was itself primitive and fragmented in very fundamental ways. Combining a basic distrust of authority (well earned) and quickly discovering errors and contradictions in what the "teachers" said or wrote, I determined to follow a pattern of research first laid out in 1637 by Rene Descartes. In his *Discours de la Methode*, Descartes gave his four rules for scientific inquiry – good then and good today. They are: (1) never accept as true anything that cannot clearly be seen as such; (2) divide difficulties into as many parts as possible; (3) seek solutions of the simplest problems first and proceed step by step to the most difficult; and, (4), review all conclusions to make sure there are no omissions. I add; "or contradictions". I then proceeded to apply these dictates to the field of "winegrowing" – that is, the entire *continuum* with no division of "importance" anywhere along the sequence.

My approach may appear to have mathematical overtones but it is "apparent" only – not of substance. There are no "equations" in the true sense presented – only notational statements for ease, simplicity and fractionalization. It is necessary in order to break down to the basic factors.

Some scientific theories depend primarily on inductive reasoning: analyzing a lot of experimental findings and then constructing theories that explain the empirical patterns. Others depend more on deductive reasoning: starting with elegant principles and postulates that are embraced as "holy" and then deducing the consequences from them. All scientists blend both approaches to differing degrees. With a good feel for experimental findings one can use this knowledge to find certain fixed points upon which one can construct a working hypothesis which, hopefully, will rise – with repetition and predictability – to the level of "THEORY". The classical methodology is fine as far as it goes but Einstein probably best described the process of discovery when he said, "A new idea comes suddenly and in a rather intuitive way. But, intuition is nothing but the outcome of earlier intellectual experience".

This heuristic unified theory of viticulture does provide us with predictability in that it tells us that if we do "this" then "that" will occur. The "prediction" aspect of any hypothesis is critical. That is, the hypothesis to be useful, must be fulfilled by the result when an action is taken. That aspect, in and of itself, constantly focuses our attention on the idea that we are dealing with a "system". If we can accomplish that mindset – by *teaching* – then we can break free of the historical pattern of "static" thought in approach to viticulture. That is, that every move we make affects the "system" down the sequence path to the glass of wine. If one looks carefully at the "studies" or "projects" of past viticultural activities one will find they are static in nature and therein lies the/a problem. The shift to a "systems analysis" mode of thought is embodied in the approach this theory propounds. Every bit of commentary noted in the vit discussions views the subject point as part of a system and approaches it in that fashion. No one aspect stands alone even though we do separate it out for analysis. The operation of the point is within the system and controlled thereby. This integration is what is important to the winegrower. The *control* of the integration is, in fact, the essential job of winegrowing. In order to *control* one must first understand the parts and their interworkings. And, I point out, the word "control" inherently includes the concept of "goal"! To what end is the control being applied? One can learn the mechanical details of the interworkings – then what? If one does not know the "goal" then control is irrelevant.

The fact is that the "system" approach has been pervasive throughout the history of viticulture but at the subliminal level – not overtly propounded. The habitual "do this, do that" practiced came, actually, from system observations of the old guard. If one did "this" the wine turned out good. If one did "that" it didn't. All we are trying to do now is explain "why" some practices were good or effective and some were not. Once we make progress in the "why" then we can change the "how" to achieve our "goal" – whatever that may be. This new viticulture is the outgrowth of the system approach within the knowledge we now possess or hypothesize. Now widely practiced, it has provided generous returns to the state of American wines. Still, with further gains in knowledge better systems may dictate themselves in the future. The system is now being used widely in the newer vineyards not only in California, Australia and Europe but also in Argentina, Uruguay and Chile. Magnificent wines are coming from these places though few of the finest are making it here. In fact, some of the most perfect applications of the system I have seen are in Argentina and Uruguay. Absolutely phenomenal wines are coming from there particularly Andeluna, Lindaflor and Azul from Argentina (Tupungato, Mendoza) and Bouza (Tennat and Tennat/Tempranillo blend) from Uruguay.

As for the effectiveness of the presented discourse I do point out that there was a time not long ago when the "experts" and the "cognocenti" asserted loudly and widely that not only were the wines of Monterey terrible but that it was *impossible* to do reds there and particularly the Cabernet complex. The Ventana has not only seen gold medals rain down for whites but also for reds over the last fifteen years – including Cabernet and blends! What happened? The climate is still cold and windy. The soil is the same. The winemaking techniques are mostly old-world and nearly the same. The only real difference is the viticulture techniques. I attract your attention to the many, many awards my neighbors in the region are now receiving for reds and whites. The difference? Their vineyards have changed to the new system. And so have they.

This treatise is intended to explain the history and evolution of the "new" system of viticulture and the basic principles behind it. It does not encompass the myriad day-to-day applications in the field except in some important structural areas. A full approach would be tedious in the extreme if all the potential scenarios were covered. Those applications are best left to in-field situational discussions. "Seeing" is far more understandable and effective than a bunch of words that deaden the mind can achieve. With the fundamentals in mind, field supervisors and middle management can sort out applications usually, although some are arcane and not readily apparent. Those I cover with clients as needed.

Perhaps twenty five years ago Ms. Zelma Long was a star winemaker on Bob Mondavi's staff. She left there and became the winemaker for Simi Winery – the first female to be head winemaker for a major winery, I believe. Later she became head honcho of Simi – another first and well-deserved. She was, and is, a talented and marvelous *winemaker*. In that roll she called me one day and asked if she could come visit and explore this nonsense I was expounding. As I recall, Simi had been using some of my wines as controls for evaluating varietal correctness. I said sure and she arrived. We spent several hours in the vineyard (can you imagine – a winemaker in the vineyard!!). At the end she said "Doug, what you are talking about is *making* the wine in the vineyard". I said "Exactly. That is precisely where superior wines *are* made!" I tell this story to emphasize the point and to demonstrate how shocking to a winemaker it was at the time. Sure – it was given lip service in myth but not in practice. My goal was to make it "all right" to *think* in the vineyard, a practice not terribly common then and certainly not expected. Growers were to do what they were told – to be automatons following Davis' and wineries' instructions to the letter. Yet – those instructions were "one size fits all" in nature propounded by people who were automatons themselves regurgitating the fixed party line – the conventional wisdom geared to industrial wines of hot country. Monterey and her cool climate presented an entirely new set of conditions requiring a new set of solutions.

For years people have been pushing me hard to write about our work and weird ideas. I have resisted because I felt there were errors and inconsistencies in my thoughts as they were hypotheses in development. The main thrusts were so contrary to conventional methods and the published academia writings (and media beliefs and myths) that I was sure the initial resistance was going to be hard and of long duration. It was important, I thought, to keep the attention on the big picture. If I wrote I would provide a "fixed target" on trivial errors. I needed time – lots of time – to "prove" the methods and demonstrate their efficacy and to sort out the errors as much as possible.

When I was maintenance test flying within an hour flight I had an answer – yes, no or maybe – and could go out and check it out the next day. Here, from hypotheses to acceptance or rejection on white grapes is about six years and eight to ten years for reds.

Now that I am no longer "in the game" I can recount the development and explain details that I have failed to mention or that have been missed by most who have incorporated the major elements. Hopefully my commentary will trigger some synapse in a young person's brain that will lead to even greater leaps forward in the quality of America's wines.

Having "passed the baton" of active participation I now am called upon for consultation mainly in two areas – procedural guidance of senior management/ownership and problem identification and solving for senior and field management. I primarily try to emphasize meaningful communication up and down the chain. Senior management is usually not well versed in the details of field activities focusing mainly on finance, marketing and macro-costing. That lacking renders them unable to conduct accurate and meaningful cost/benefit analysis for ownership. Field management normally is not equipped to tender thorough analysis and usually not provided with all the considerations when asked to do so. Neither side is particularly fond of admitting weaknesses and will go to great lengths to blur the matter. Sometimes an outsider just provides a fresh set of eyes and a mind with no axe to grind. When so occupied my goal is to help – not find fault in a picayunish way.

I thank God for the life I have been allowed to live and to live it in this time in history.

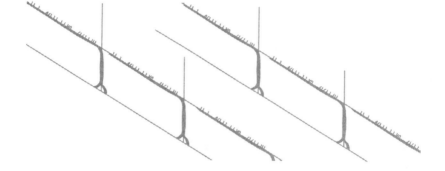

Reason, Observation and Experience
The Holy Trinity of Science
—T. H. Huxley

Section 1
INTRODUCTION

Over the last twenty-five years winegrowing has radically changed from what was before – not only in California but in many places elsewhere in the world. That major change is still underway. The focus and purpose of this text is to put the evolution of this system into its historical context of the time, provide the rationale of stages and elements and the details so important to its maximum operation. While we see the widespread application of the system there is also a widespread lack of understanding of the reasons "why" which lead to mis-application. It is hoped that comprehending the evolutionary details and their raison d'etre will tickle the brains of young people into the future gains in quality of American winegrowing.

Such a quest by its nature is unending. So many more things remain undiscovered or not yet understood. Actually, there is immense joy (and frustration) in the "Quest".

Once a person's curiosity, on any subject, is aroused it is surprising just how far it may lead him in pursuit of its object, how readily it overcomes every obstacle. In my own case my absolute fascination with understanding the forces at work while faced with the "new" factors of cold climate and wind has consumed me for the last thirty-five years.

Throughout the text I will discuss detailed historical aspects and conditions pertinent to the vitcultural subject at hand but first I will present herein general California historical conditions leading up to the evolution and development of the "new" system and its need.

Also throughout the text I will be presenting my perceptions and interpretations of situations. They are mine alone and they are just that – perceptions I have formulated. They may please some and they may anger some. So be it. Actually, those they anger may like to discuss those angers with me and the reasons therefore. It would be a fun conversation!

In those early days I was the classic "outsider". I am not a Californian, my last name does not end in a vowel and I was never a student at U.C. Davis or Fresno State. So I will begin with my background to let you know what I am – not what I am not.

I was born in Seattle, Washington in 1940. My mother was one of nine children of a German farming family in Eastern Washington – farming wheat, cattle, and hay. From the time I was a very small boy I would spend my summers working on the farm of one of her brothers. As you may surmise, there was this weird idea that male children had chores and that they worked in the fields with the men. At seven years old I was driving tractor (an old Ford-Ferguson) hauling hay. It was the old world way and I loved it. For a boy to be with the men was fantastic – every boy's wish. The work ethic was up before dawn, certain chores and then a massive breakfast of eggs, potatoes, fresh baked bread (the women were up long before the men) slathered with sour cream butter (if you've never had sour-cream butter you've missed one of life's joys) steak, bacon and pancakes. We ate like anaconda snakes except we wouldn't lay around after – it was to the fields for hard work until noon. Then – another large meal and back to work. It was great.

In 1950 my mother married my step-father and we moved to his home in Wenatchee, Washington. It is dead-center in the state located on the Columbia River. It calls itself "The Apple Capital of the World" – perhaps a little presumptuous. My step-father had served with Patton in Europe in World War ll. When the war was over he returned to Wenatchee, became a postal clerk and owned a small apple and pear orchard. His various relatives owned apple orchards around us. Ah, yes – there were the chores for boys. Shoveling snow from the driveway was one – and one of the reasons I now live in California! From the seventh grade on I played football. Apples in those days were selectively hand-picked into boxes and stacked in the orchard by pickers as they filled the boxes. They were subsequently loaded up by hand onto trailers and taken to warehouses where they were cold-stored and processed. After football practice and walking home (about 4 miles) I then "swamped" apple boxes out of the orchard often until midnight. Boy, was I always glad to see apple harvest finish! Each summer I still went to one of my mother's brothers' ranches to work the summer. During spring school break I would also go for spring planting of wheat.

Starting at twelve, I had what was called a harvest permit or license to drive trucks loaded with wheat from the ranch to the grain elevators in town during wheat harvest season. By this age I was also trained on big crawler tractors. I followed this pattern through high school. When self-propelled grain harvesting machines (called "combines") were developed I was sent by a non-family employer to John Deere combine school to learn the operation and maintenance of these new machines.

Upon graduation from high school I crossed the mountains to Seattle to be a student at the University of Washington. This country boy had no idea of what to study other than mathematics. I just knew I needed a college degree for the Navy. I knew I wanted career Navy and I was in NROTC. At the end of my freshman year I dropped out of NROTC and became a "naval science student". I saw that I didn't want to leave university in four years, that I wanted grad work, and that I couldn't afford all the drill time and navy activities time as I was working my way through school. Thus, I could attend all the Navy classes except classified ones and had no drills. It also meant I would not automatically be a commissioned officer upon graduation but would have to go through an officer training program at some point in the future if accepted at the time.

I continued to putz along in Math and Economics. In my second year I took a little time off from school and surveyed for Great Northern Railway – now Burlington Northern Santa Fe. There I learned some skills of great value later in life. Then, back to school. In my junior year I got a menial job part-time with a company called Larry Smith and Co. This was of immense value to me later in life.

Smith and Co. was a real estate planning and analysis group. They did the work on the first shopping center development in the world – Northgate Center in Seattle. Of course, that led them to do the work on the subsequent ones for a lot of years – Lloyd's Center in Portland, Hillsborough in the Bay Area and all the others in that area and elsewhere in the U.S. That work led them into city planning. Vancouver, British Columbia, is built today as I and another fellow sketched it out on a bar napkin in San Francisco! Happily – we kept the napkin for the next day.

All these analysts at Smith had masters' degrees in a wide variety of subjects and many had attended East Coast Ivy-League schools. They were a very sophisticated group. At Smith the pattern of work for these professionals was often into the midnight hours. My job was cross-checking their work so I was there working as well. Partners and analysts would regularly go out to a fine restaurant for dinner during which they discussed projects, world affairs and economic theories concerning city expansions or re-developments. I was directed to join them. Rather than laugh at my naiveté they made it a project to civilize me in the ways of food and wine. They would often each order a different food and then each would have me taste theirs and explain its name and preparation. It was often a competition among them to see who could dazzle me the most. It was the same goal with wines – French – and who could delight me the most.

Of course, somebody's explanation would engender a disagreement about "proper" food preparation or "correct" ingredients or the relative merits of a wine. How lucky I was and you can see why I say this was of immense importance to me given where the whims of the gods eventually led me. It was an incredible opportunity for a young lad whose frontal cortex was still developing to be immersed for three years in such an intellectual environment. Within a short period of time I was moved from my menial tasks into positions of much greater responsibility even though still attending the U. Eventually we moved the West Coast office to San Francisco (which move I was in charge of) and for nearly a year I commuted by air between San Fran for work and Seattle for school. Life was good and San Fran provided a much greater world for the food and wine games than did the Seattle of the time.

While I could have had a career with Smith and Co. (after military duty) I was still focused on Navy eventually. I decided I wanted experience in industrial engineering and manpower analysis. I sought and received employment with Boeing as an analyst in manpower loading. I had eight "shops" under my control in the commercial division – this was about 2500 people or so. The work involved projecting manpower requirements by training level against aircraft orders and projected orders and the training lead times of personnel needed to accomplish the production. It was fascinating work in the beginning when learning the process but it wasn't too long that the repetition became boring. It was where I also learned that all the rules, regs, limits and office politics of corporations were not for me. It was eventually the most mind-numbing experience I ever had. Stay within the lines, make no waves and put in your time was the mantra and attitude. It was horrible for me. For others it was comfortable.

Throughout all this time we had a little thing in America called "The Draft". I was classified I-A, a picture of perfect health and I was single – absolutely prime meat for the grinder. There was some waiver for college students, depending upon the Draft Board's needs, but I was also getting beyond the four year allowance by my Board. I was aware of a little quirk in the system. When one applied for Naval Air a letter immediately went to one's Board deferring one for consideration temporarily until the Navy made its decision. It was the only program like that because of the Navy's special needs in that area. Probably because the Navy needed people of intellect and athletic skill who were also crazy as hell to want to fly off boats!! There weren't many accepted and fewer yet who successfully completed the program.

When I would get a draft notice I would apply for Naval Air and get deferred. I then took all the tests and eventually the Navy's physical. When it came time to say "yes" I said "no". It would take about six months for paperwork delays then the process repeated.

In 1965, I was approaching twenty five years old and the upper limit for entrance to flight school. It was becoming time for me to get on with my real life – a naval officer. I had my degrees and my experiences in the civilian world, knew I hated the corporate world and that Vietnam was heating up. I thought I belonged wherever America was at war. I was a child of World War II and was raised to believe that we each had a duty to this country that our forebears had fought for. My ancestors had fought in the Revolutionary War, The Civil War, The Spanish – American War, WWI and WWII. Now it was my turn and duty. I would become a "Knight Templar" and save America from the Barbarians.

In September, 1965, I entered the Navy in the Aviation Officer Candidate School system. I finished that four month school and was commissioned an ensign in the U.S. Navy. I entered Primary Flight School at Pensacola, Florida. I was then sent to Basic Jet School at Meridian, Mississippi in early to mid 1966. Later in the year I was sent on to the Advanced Jet School outside Corpus Christi, Texas to fly the F9-F Cougar – a post – Korean War swept-wing fighter that was still used as a trainer. In the Navy, one must master instrument flying and Carrier qualify before receiving one's wings. In 1967 I finished up the training command first in my class, received my wings of gold and chose attack-fighters as Vietnam was an attack war not an air-superiority type war. As 'first' I received my choice of aircraft and was sent next to Naval Air Station Lemoore, California to the Replacement Air Group (RAG) to learn to fly the A-4 Skyhawk and to learn advanced fleet tactics. In late 1967 I finished the RAG mastering navigation, bombing, strafing, nuclear delivery, carrier landing and a wide variety of other trivia. I was assigned to my fleet squadron- VA212 – which had just returned from Vietnam and was scheduled for a very short turn-around back to Vietnam.

In very early 1968 we were on our way aboard the U.S.S. Bon Homme Richard – a WWII carrier converted to angle deck. It was to be my home for most of 1968. During that year I had my parents acquire a parcel of raw land adjacent to their small new orchard for planting to apples eventually. My intent was to have orchards to retire to post – Navy if I survived. If I didn't my parents could have it. In those days the running joke was "The definition of an optimist was an A-4 pilot assigned to Vietnam who gave up smoking for his health".

It was difficult working by mail with an irrigation company to do what I wanted done on that land as it hadn't been done there before. When I returned from Vietnam, I worked with them face to face and thus we were able to install the first over-tree permanent-set irrigation system in apples. Both they and my parents thought I was nuts. Of course, it is now the norm – whether over – tree or under – tree. The orchard design was also new but not the first – using semi-dwarf trees at much closer spacing than the historical way. I'll return to this later in technical discussions.

It was also during the 1968 visit to Vietnam that I provided "seed" money to a school friend of mine for a "prospectus" on development for a "project". Let me here digress into some background information. At that time the U.S.

tax structure was essentially confiscatory in nature. Passive income was taxed at the 70% rate. Thus the development of "shelters" arose and the "limited partnership" form became the vehicle. Without putting too fine a point on it, and very simplified, the scheme was essentially to put up as the required "General Partner" an empty pocket entity. The "limited partner"/investor, for example, put up a $1,000 investment and the partnership bought ground with a $4,000 insurance company mortgage. The 'investor" would then, over a short time, receive a $5,000 tax write-off!

If the project made good the General Partner did well with their high ownership position in addition to their on-going "management fee" all based on zero cash inputs and the limited partners made out with all their previous tax write-off and sheltering for a few years plus they now would have positive income flow on an investment of which they had long ago more than recovered.

If the project failed the General Partner had empty pockets. The limited partners had already made good returns based upon their write-offs thus sheltering their other income. And – as California is a "sole recourse" state which means that if you lend money in the purchase of real estate the "sole recourse" of the lender in event of default is the real estate in question *itself.* Therefore the land would become the property of the insurance company by foreclosure, the GP would just go away having earned nice management fees but with no liabilities and the LPs would have their return with no liabilities. The banks and insurance companies would get to clean up the subsequent mess.

Now, this is an oversimplification but that is basically how it worked. The "seed" money I had provided in 1968 was for the "prospectus" on such an operation that involved acquiring land and the planting of Christmas Trees. There were (and are) some very interesting tax elements to "timber" farming that come in to play. The project was formed, the installation accomplished, and everyone was happy. Years later it failed and the above noted scenario played out.

Meanwhile, the main promoters moved next into grape pro-motions – formulating plans for vineyards under the same approach and acquiring a "nursery" operation. My original "seed" money tied me into an ownership position in any future projects with no further cash inputs.

Jumping back, by mid 1969 I was on my way back to Vietnam aboard the U.S.S. Hancock – another W.W.II carrier conversion similar to the "Bonnie Dick". I was occupied there until mid 1970. I finished my combat time with 329 combat missions and was assigned to Naval Air Station Lemoore, California as a weapons and tactics instructor in the A-7 RAG. The A-7 was the new modernized replacement aircraft for the ageing A-4. I was to have a two-year tour of duty there – it was the "cream" duty for a career pilot at the time. I thought not. In the Navy I had met some of the finest men to be associated with you could imagine. At the same time I had met many of the opposite color. I came back with a long brewing anger at just about everything – the incredible stupidity of the Vietnam war, the incredible stupidity of senior officers' prosecution of the war, the waste of lives, etc, etc.. In essence I was "combat sick" and had acquired a strong dis-respect for authority. That is not a good thing for a career military officer. I came to the conclusion that I had done my share for "democracy" and it was time for me to promote myself to civilian. The Navy was not for me. I would go home and become an "Apple Baron", sequestering myself in the peace and quiet of trees and mountains. Earlier, after acquiring the orchard land, I had immersed myself somewhat in reading information in apple publications. At this decision point I heavily dug out information from the works being done in England, Holland, France and Australia in addition to the U.S.

Here I will digress into a discussion on apples as it has significant and germane bearing on later grape analysis discussions. You might also note that this was thirty-seven years ago – before the internet- and the acquisition of information was slower and more tedious than nowadays!

When I was a boy a "standard" apple orchard was planted on "40 foot centers" (SEE PLATE 63). There were 40 feet between the rows and 40 feet between the trees within the row – *at* maturity the space needed. "Maturity" took 15 – 20 years. When I was twelve my step-father and his brother developed a new large orchard and they followed the 40 foot layout. However, they planted peach trees at 10 foot spacing among the apples. They explained to me that peach trees develop much much faster, that they are smaller trees, that in early years yield is a function of density and that as tree size increased they would pull out peach trees as necessary. Wow – that was clever, I thought. They would be peach growers until they could be apple growers again – eventually. Another grower nearby did the same thing only a little differently. He planted high density but of only apples. Again, he explained to me the relationship of yield and density in early years and the subsequent removal of trees as necessary to provide space for the ultimate 40 foot spacing needed at maturity.

In those days a mature orchard was characterized by a big tree with a globe-like top with a layer of foliage over the globe. The trees were tall requiring tall ladders enabling pickers to reach the fruit to harvest or, in the spring, to thin the fruit. The pickers carried bags on shoulder straps up and down those tall ladders and moved the ladders around the tree. To further compound the issue was the "Common Delicious" apple – the prevailing strain that ripened or colored un-evenly. The money was in a grade called "extra fancy" which was color determined. This resulted in a practice of selective picking and, thus, multiple harvesting passes through the orchard. It was not until later that more evenly coloring strains were developed and better growing procedures came along. A good grower in those days would get about 200 bushels or so per acre in a decent year.

Apples were a crop of importance in America. Johnny Appleseed was known by every school child. "An apple a day keeps the doctor away" was known by everyone. "Mom, the American Flag and apple pie" was known to all. I think every land-grant college in America had a department of Pomology – the study of apples. Apples were harvested in the fall and stored in cold-storage warehouses but by January even the best-storing varieties were starting to become "pithy" in the centers, by February they were essentially gone. Only canned or preserved apples were available until the early apples of July.

All the inefficiencies, some noted above, of the apple business combined with human fondness for apples had led to tremendous research efforts not only throughout America but also those countries noted above in addition to others. State and Federal funds flowed into those projects. In America can you imagine a politician voting *against* funds for Johnny Appleseed? For apple pie? It would have been political suicide. Compare that thought to voting *for* funds to research alcohol containing beverages such as wine in prohibition-oriented America. Talk about political suicide! The attitude toward wine at that time was that it was mostly used by "winos" – alcoholics drinking the fortified concoction or by ethnic communities of the Catholic persuasion both of which had some difficulties with general public image.

Vic Allison, a pioneer of the Washington wine industry, was elected to the school board in Grandview. When it became known that he was an owner of a winery a recall election was held and he was removed from the board – all because of wine! Such were the times and attitudes.

By 1968 a major revolution in the western world-wide apple industry was occurring. Not only had better more uniformly coloring strains been developed but a huge paradigm shift in plant and orchard design had occurred. The Merton-Malling and East Malling research stations in England had developed dwarfing and semi-dwarfing rootstocks in addition to those in the U.S. The Dutch were also well along in this area with their "meadow apple" approach, The studies of light penetration and the action of sunlight upon bud fruitfulness as well as its action on coloring had been determined. The relationship between structural design, exposed leaf surface area, photosynthesis substrate production and its mathematical expression had been published. The relationship of oxygen, calcium, ethylene gas and the deterioration of stored apples had been elucidated. The development and installation of "controlled atmosphere" storage allowing nearly year around fresh apples was underway.

Various planting schematics were occurring. Like all things there were pluses and minuses to the myriad systems being tried. The full-dwarf rootstocks were shallow rooters and susceptible to trees falling over when full of apples or subjected to winds. They required fancy trellis systems that were expensive. But, in certain areas and for certain purposes they were desirable. Cutting to the chase and skipping the pros and cons as non-germane for this writing, I chose semi-dwarf rootstocks for my orchard in 1968 (one had to order trees from a nursery a year in advance). My orchard was laid out at 18 feet between rows and 12 feet between trees in the row. The semi-dwarf rootstocks would root well thus no expensive trellis would be required. The training of the trees was such that light penetration channels throughout the tree were maintained. The design was such that roughly 8 to 10 times more light exposed leaf area per acre over the old domed standard tree was achieved. In mathematical terms, grams of fruit per hundred square centimeters of exposed leaf surface was a new design and projection tool. The smaller, shorter trees led to easier thinning and harvest (read "cheaper") or, at least, less expensive. By the time these modern orchards came into maturity a good grower achieved not 200 or so bushels per acre but varied between 1200 and 1500 bushels per acre of much higher quality fruit! *That* is what I mean by a *revolution*. The results of this design? As a member of a huge cooperative, we received the "Top Grower" trophy for 1977 and 1978! The quality was off the charts and the yield was nearly double surrounding orchards- all because of design change!

It didn't occur on the front pages of the newspapers – but it was a revolution none-the-less. Light, Light, Light! The development of the new trees allowed the new light knowledge to be applied. We will return to this.

Another area from my background, of lesser importance but still affecting my thought processes, was that of grain growing. The land-grant colleges throughout America had studied carefully plant spacings for different soils and areas for grains, soybeans, corn, etc. There were many variables considered including regional rainfall patterns both annual and seasonal in accord with crop patterns or, if used, irrigation systems and water availability. Different dry-land farming situations resulted in different planting densities. As a boy when I was planting wheat I would have to adjust the drill differently for different fields and in one field my uncle had me adjusting it differently for two sides – one clayish and one more sandy. Thus, you see that the density business was instilled in my mind from early on in agricultural endeavors not grapes. Further, the academic work in Economics and Mathematics formulated a way of thinking with equations and graphs which was applicable later on with grapes.

Let me now return to 1970 and my stationing at Lemoore, California which was not too far down the road from Madera, California. As I mentioned, the promoters had acquired a grapevine nursery which was located in Madera. They had had a state nursery inspection and had had their license suspended for an outbreak of "crown gall" (or Agrobacterium Tumafaciens). They didn't know what to do about it and wanted me to look into it. There was no literature on this situation and the state nursery service had no advice.

I quickly found that this rascal plagued the rose industry and that they dipped the *dormant* hard wood rose plants, once dug up, in a solution of Clorox and water. The easy part was washing down the facilities and benches and installing "step pans" at each door all with Clorox solutions. The hard part was finding a solution that would kill crown gall but not damage the tender green snippets of grape vine material that needed dipping. However, find it I did. The nursery was cleaned up and the license reinstated all in about 60 days. It was a relief to all. A curious aspect was that the nursery service asked for the procedures I followed to help others as needed. I gave them the steps and solution ratios. About a year later they let me know that they "thought the system *wouldn't* work but to keep doing what we were doing because we were clean"! Amazing! There never was an outbreak again in this nursery though others in the state did have significant problems with crown-galled vines over the following years. I returned to flying and apples as my foes.

Let me now digress into a discussion on grape vines and the development of so-called (at the time) "Super Clones". At the time the wisdom was that there were 13 "known" viruses that attacked vinifera grapevines. These viruses would not bother thee and me but were specific to grapevines. They affected grapevines in various ways, some devastatingly, some with less impact but all have had economic impact. Some severely damaged a leaf's ability to conduct photosynthesis and production of substrate. Some damaged the fruit itself. Some damaged the vascular structure. Nasty, nasty little things. There was a commonly held belief that I heard uttered often that all grapevines in the World were infected with virus to some degree or another. I don't know if that is true. I doubt it – but it doesn't really matter. Viruses were of significant and severe economic impact in terms of yield and certainly negatively affected wine quality – although in some instances they may have enhanced wine quality! Consider a virus that reduced berry size in an otherwise large berry producer. The skin to juice ratio would be greater and thus more extract in the wine. One has to respect the law of unintended consequences.

If memory serves, I believe it was in 1966 that Dr. Austin Goheen at U.C. Davis perfected and introduced the "Heat Treating" process to eliminate known viruses from grapevines. Goheen selected varietal clones from the "Foundation Plant Material Service" (FPMS) and from the "Vit Block". The

FPMS is our repository of plant material jointly operated by U.C. Davis and the state nursery service. It is the source of all state approved grapevine material for propagation. The "Vit Block" was a block maintained by Goheen and Davis for use as comparatives for varietal identification. Much of it was acquired by Goheen and others from schools in Europe. Its wood was not available for distribution until it had gone through the FPMS process. Subsequently, Goheen controlled the release process and he would allow no material to be released until it tested clean of virus within the testing limits of the time. Prior to this process the selections maintained by the FPMS and Davis were the result of a completely different process.

I have perused writings written throughout the more than one hundred last years and there are two, and only two, criteria mentioned for the selection of wood for propagation. They are (1) yield and (2) *apparent* disease – freeness! In American writings back to the turn of the last century I have found not one reference to wine quality! Now – that does *not* mean that winegrowers did not consider such a factor within a variety. It only means that that factor was not of sufficient interest to the scribes nor was it of sufficient interest to the academics who wrote university extension service instructions or guides to growers. Selecting clones for wine quality did not come along until much later.

Simplified, Goheen selected those clones of varietals, grew them in pots and subjected them to "heat chambers" of 105 degrees Fahrenheit. The heat did not kill the virus in the subject plant. The vine grew very slowly at that temperature and the heat held back the virus from the new tender tip. He then snipped the "tip" and grew that into a plant, tested for virus and re-did the process until the selected plant tested (within the accuracy of the testing procedures of the time) free of "known viruses". Such a plant became a "Mother plant, assigned a clone number and plant material propagated there from was made available from the FPMS, both dormant wood and green material. This material was made available first to licensed nurseries and then to growers, if available. The desire was, rightfully, to expand the availability of these new disease-free plants to the industry as fast as possible without waiting years and years for hard wood growth.

Nurseries purchased "Mother" plants, grew them in greenhouses and, as they grew, cut off "runners" (tender canes). These they would then cut into one-bud sections with their leaf, place them upright in a bed of perlite, and root them under an irrigation system of "misters" that misted the beds every minute or so. When rooted they would be transplanted into pots of growing material. When developed they would be taken from the greenhouse, "hardened" off and shipped to growers for field planting. This process was called, cleverly enough, "Mist Propagation" and was a very rapid method of expanding material. *One* plant could theoretically become a million within one year! You might note that the roots under these vines were their own – that is, vinifera.

Goheen constantly insisted on referring to these plants as "known virus free". However, in the hands of tax-shelters promoters and the general media these plants became known as "Super Clones" and because free of debilitating diseases, would grow at super rates and produce big crops early and forever of mostly superior fruit! A miracle!

Today, now that the euphoria has long since subsided, plants from Goheen's heat-treating process are the foundation of our approved plant material supply. The FPMS has diligently improved the early testing methods and greatly expanded both varietal and clonal availability to the industry. Selections have moved beyond the simple selection criteria of "disease freeness" to areas of wine quality generated and climatic appropriateness, to mention a couple.

During the remainder of 1970 and through 1971 the tax-shelter promoters I had helped fund were busy not only with the "tree farm" and nursery but they also were busy putting together vineyard limited partnership projects. Periodically they would ask me to come look at a piece of land they were considering. Except for that I was focusing on flying and apples. They had become entwined with Ed Mirassou and he, of course, steered them to Monterey.

My tour of duty at Lemoore was to expire in mid to late 1972 and I was planning on resigning from active duty. I was often flying to Wenatchee on weekends for apple work. By late 1971 it appeared to many that the war in Vietnam was starting to wind down. The Navy began to consider reducing forces and in November, 1971 Admiral Zumwalt (Chief of Naval Operations) sent a message that anyone who wanted out could get out if they did so by December 31, 1971. I left active duty on that date.

The promoters meanwhile had been active. In 1970 the Bank of America had released a study that predicted that the wine industry would grow over 8% per year for the next decade. They had employed top-flight San Francisco lawyers and accountants and financial managers. They had Ed Mirassou (a *very* prominent man in the industry then) and they wined and dined many professors. They were on a roll.

Their first projects were to commence early in 1972 involving a 1350 acre piece of land of which about 1100 acres were plantable divided among three limited partnerships. They had several more projects in the works involving other lands and partnerships that would, within the next couple of years, total out to nine limited partnerships encompassing about 2500 planted acres. I was a small owner in all.

They heard from others that I had left the Navy. They contacted me and asked that I clear the land, drill wells, install the trellis and irrigation systems and plant the vines. In other words, do the development. After much discussion I agreed to give them three years of development time at an increase in my percentage ownership plus salary plus all the time I needed for apples. And so – I came to Monterey in the late spring of 1972.

The only thing I knew about a grapevine then was that I thought it was probably a good idea to plant them green side up. I knew nothing about farming winegrapes nor did I know anything about winemaking. My work for three years was to be industrial engineering in nature – time, materials and manpower control – and then I was going home. Little did I realize that I *was* home for good!

Those who cannot remember the past are condemned to repeat it.
—George Santayana

Section 2
MAGNIFICENT MONTEREY

Prior to the planting of Monterey, Napa was considered the "cool" climate area of grape growing. The primary focus (almost the whole focus!) of university grape study was the Central Valley of California. Fine wine production in America was trivial in quantity and, in most cases, in my opinion, also in quality. When I speak of poor quality of California wines historically some, I'm sure, will take offense and curse me as an old geezer whose memory has faded or has some axe to grind. Well, in defense of my observations I tender a couple of references. The first I attract your attention to is the Davis 20 point scale evaluation, a form almost universally used until Parker's 100 point scale came down the pike (by the way, there were others using a 100 point scale before Parker – both here and in Europe). You can probably pull this form up on the internet or find it in a book (for you young people – that is paper with printing bound together like this tome in your hands). Look *very* carefully at the categories and the assignable point ranges. Volatile acidity gets a 2 point spread! Today, if a wine reeks of VA it gets a total zero. Look at "clarity". The categories themselves and their point ranges indicate the qualities of the extant wines.

A second reference you probably won't believe so go to page 711 (records and scoring) of "Table Wines" 1970 edition. You will find "...widespread indifference to critical sensory examination of wine in the industry. Our experience is that this is true in Europe also."! In short – wines were not critically *tasted* upon completion! And it sure showed in the bottle! Remember – this book was the bible of the industry at that time and was the teaching text for fermentation science. America had not yet evolved into the level of lifestyle it is today. The empirically acquired, accumulated knowledge of vinifera viticulture was "hot" climate knowledge of the central valley geared to "industrial" wine. Notice the locations of the schools! Any visitor to St. Helena or Calistoga in Napa Valley in August would be hard-pressed to call it "cool" unless, of course, they had just driven in from Fresno or Bakersfield!

Monterey is, in my opinion, the coldest winegrape growing region in the United States during the growing season. The application of warm climate techniques combined with wind, overhead irrigation, absentee ownership etc. would lead to a disastrous time for Monterey during its birth.

Let me here digress into a discussion about the reference word "REGION". The regional concept of viewing winegrape country was developed at U.C. Davis (Winkler, et al) as a tool to evaluate the potential of an area and as a tool for recommending grape varieties. The method was quite simple. It was believed that at around 50 degrees Fahrenheit a grapevine commenced its activity. Therefore, one simply took the peak temperature of the day and subtracted 50 generating a number called "heat units"! These daily heat units were summed up over a time period roughly amounting to the growing season. This summation resulted normally in a number between, say, 1500 to 4000. The process could be followed for, for example, Bordeaux, Burgundy, Chablis, etc., etc.. Temperatures were reported and recorded in that fashion – highs and lows of the day. Once the summations were calculated comparisons could be made between regions of the World – or so it was thought. The coldest areas were called REGION I and, by 500 heat unit spacings, reached Region V for the hottest. By looking at the varieties successfully grown in Europe within regional temperature classifications they could predict which varietals would do well where (see General Viticulture, 1964).

U.C. Davis did that work and disseminated it for much of California. However, in the process they made an inherent assumption that the day everywhere followed the Central Valley pattern- the Sun came up, there was a steady gain in temperature to a maximum which held on for awhile and then a slow decline late in the day. That pattern simply isn't the case in many areas and certainly not in Monterey and in many of the coastal regions now planted to grapes! Those areas are much cooler than the system indicates.

It was a fine, simple procedure within the data limits of the time and, in many instances, worked just fine. But, like many "simple" things, there were (and are) some devils lurking in the details – devils that would have substantial impact in Monterey and lead me into other areas of analysis.

The Regional System indicated that Monterey ranged from I to III mostly falling in the II to III zone while some small southern parts could hit IV.

By mid 1972 I had reached the point of land development that some vines could be planted. They were. I was trying to grow grapevines in a situation where they were fogged-in until 9:00 am or so, the sun would burn through and the temperature climb and then at 11:00 or 12:00 the wind off Monterey Bay would commence and the temp would drop precipitously. See Diagrams A and B on the next page.

Considering that photosynthesis is a chemical process and that the *rate* of said processes doubles with each 10 degree centigrade increase in temperature (at least in the relevant area) then the daily production of photosynthesis substrate must be a function of *average* (simplified) temperature *not peak* temperature. In fact, the relevant information is the calculus of the area under the curve above 50 F not the arithmetic peak minus 50.

Mirassou had circular temperature recording devices that inked temps on a graph continually on both their vineyards east and west side of the valley. I acquired those graphs for the past several years and re-did the calculations on an hourly basis. That work showed that we were from one to

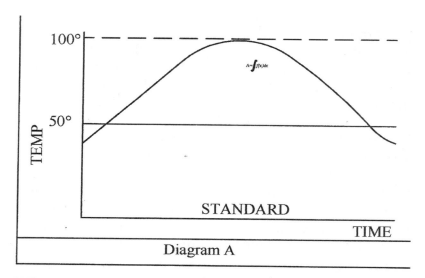

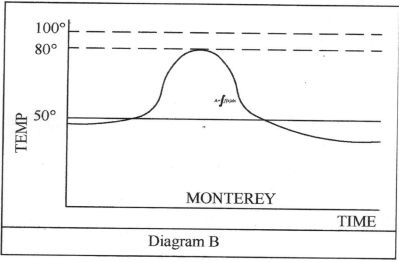

one and a half regions colder than indicated by the conventional system! At times that put us in REGION 0 – if there was such a thing, which there wasn't!

An interesting side question is why is Monterey so cool lying to the south of much warmer areas as it does? Remember, our latitude is roughly through Northern Algeria! In my opinion it is the combination of the huge size of the Central Valley and the Japanese Current, the flows of which had recently left Alaska. During the summer months the Sun rises and begins to heat the air. As the air heats it rises and creates a lower pressure area over the wide area of the hot Central Valley. Nature hates a vacuum, as they say, and lying just off the coastline is the heavy cold air lying on that cold water of the Japanese Current waiting to surge inland to correct that pressure differential. Thus the daily summer winds commence around 10 or 11 o'clock and subside in the evening as the Central Valley is no longer heating. In the winter the Valley is no longer heating up so there are no heat-pump winds through the coastal valleys. Later in the text the winds will come up again and again. Viticulturally, I solved the wind problem for the short-run on baby vines by introducing the use of Sudax (a tall growing grass) down the center of the rows providing wind breaks for the baby vines. Others soon followed the practice. This was in the time before growth-tubes were around which, when introduced, quickly became the norm given their merits. They protected from the wind, increased temperature, contained moisture and provided ease of weed control. In our area they cut one year from vine establishment norms of the time.

They were first invented in England as protection for young oak trees from deer. A fellow at Beringer brought in a few as a trial and within a few weeks ordered 20,000 more I was told. They were expensive, Quiedan (McNamara) began making them here at far less expense and then developed the two-part tube. Now they are widely available and were an important contribution.

Monterey's first little burst of commercial planting occurred in the mid-sixties by Paul Masson, Mirassou and Wente. Urban pressures were at work in their areas and they came south based upon the academics' prognostications. There was then a long pause until the tax-shelter aspect took root. The big surge of planting jumped Monterey's planted acreage from about 2000 acres, roughly, to about 30,000 acres by the end of 1974 – all within about a four year span.

The scenario of this huge exercise was: tax-shelter pro-moters/owners/manager: absentee limited partner owners to whom financial accounting of performances had to be given: overhead sprinkler irrigation (the first in America on grape vines was at Mirassou's mission ranch with heliarced aluminum pipe) that gave growers the god-like ability to make "rain": heat-treated vines that had not existed before: cold and windy climate in which there was no experience: management and staffing by personnel who had no experi-ence with grapes and no idea of winemaking or the wine-maker's needs; financial institutions requiring reports and performances and involvements: regional conformity to U.S. Davis dictates: and the list goes on. Does this sound like the opening scene of Macbeth? The witches were certainly stir-ring the pot and disasters were waiting in the wings!

By 1975 most projects were well behind projected devel-opment, financial difficulties were being encountered, wines from young vines were showing strange vegetal tones that came to be derogatorily called "the Monterey veggies" and then came a major hammer. The federal government de-cided in October, 1975 it was tired of the tax shelter write-off game and enacted the "at-risk capital" concept and made it retroactive to the start of the year. This concept dictated that the only monies one could write off were monies that one stood the risk of losing. That is, one could not write off borrowed money if one was not personally liable for repay-ment. Before this investors made okay money even if a proj-ect failed – they were farming the tax-man not winegrapes! I was told that one of the better shelters was drilling for oil and hitting a dry hole – and if you accidentally hit oil so much the better!

The "at-risk capital" change combined with the other fac-tors shut off the money faucet. The flow stopped. Over those few years management companies went under (including the one I worked for), insurance companies foreclosed and "mothballed" some vineyards and tore out others, landlords foreclosed and returned some properties to row-cropping, etc.

It was not a fun time. All the hubris and euphoria of the boom had evaporated. General gloom had set in with all its demoralizing force.

Fortunately, or unfortunately for me, I had already been infected with the grape virus. Returning home at the end of my three-year commitment was no longer being considered. I was absolutely fascinated by the problems we faced in Monterey. Though the region was damned by many I thought it was potentially one of the great wine regions of the world if we humans could learn how to work within her parameters. Sorting out those "parameters" and their inter-workings with the physiology of the viniferous grapevine became my life work for the next third of a century.

The property that is now the Ventana Vineyard had been planted in 1974 with substandard potted vines in the fall. By early 1975 it was a weedy mess as cash-flows were nearly non-existent. Lease payments were not made and later in the year the landowner foreclosed. I was no longer employed by the promoters having resigned earlier. I truly loved that piece of ground and thought it could be a world-class-vineyard someday if handled correctly. My "belief" in the region and in that property particularly was not some mystical, un-reasoned "wish" for it to be true. My research and study was already well advanced having read everything I could get my hands on that had been published in English (U.S. and England) and French. I had even worked my way through a few of the German writings. I had plowed deeply into text-books on general plant physiology. My immersion in all the "new" theoretical work being done in apples (particularly relating to sunlight) conditioned my mind to view viticulture from a perspective different from the formal, conventional California approach extant at the time.

At that very time of gloom and doom I approached that landowner to discuss the possibility of me acquiring the property. We discussed it for some time. He thought I was a "good boy". We slowly worked out a very good deal for me while I farmed the property that winter, re-planting with cuttings all the failed vines from the previous situations. We finished all the paperwork on April 1, 1976 and it was now "mine".

Many people told me I was crazy – including my parents – and they had good reason to be worried. However, I was intellectually and emotionally positive that I could mid-wife the birth of this great region-to-be.

The real adventure began!

Basic research is what I am doing when I don't know what I am doing.
—Werner von Braun

It is dangerous to be right in matters on which the established authorities are wrong.

—Voltaire

Section 3
VITICULTURE

I will take each major area, as I construe them, discuss "facts" along with my observations and opinions within the time frame then at the end bring them together. So – let's start with the grapevine and then go to the bottom of the vine and sort-of work up. It's as good a place as any.

If you steal from one author, its plagiarism:
If you steal from many, its research.
—Alva Johnston

Section 3A
THE VINIFERA GRAPEVINE

The vinifera grapevine is the wine grape group of Europe and the Middle East. Some hypothesize that it originated in the Caucasus Mountains. It is a temperate zone broad-leaf plant requiring dormancy as part of its cycle. Phylloxera is a North American pest and thus the vinifera evolved in its absence and therefore has no resistance.

Its dispersal was essentially east and west staying within temperate zones. A host of varieties evolved, many or most of the current ones probably aided by man by propagating for some purpose valued at the time. A desired plant was easily preserved by simply cutting a piece of dormant hardwood and sticking it in the ground. It would form roots and grow. Hardwood was easy to carry by people moving from one area to another to re-establish their vines. In many areas raw water was not potable and sanitation was unknown sort-of. Long before the existence of knowledge about bacteria and such, mothers noted that children who drank water often sickened or died. Drinking water mixed with wine yielded healthy people. Generals, too, noted that healthy soldiers won more battles than sick or dieing local water drinkers. Grapevines and their wine were important life-preserving tools to the ancients. The vestiges of that reverence can still be seen in some modern religious ceremonies. One medicinal practice common by the 1300's was to wash wounds with wine before sewing or wrapping (though not used by academic followers of Galen). As people moved about their vines needed to adapt to the changing locations. Vines had evolved mechanisms for this long before by utilizing bird and animal digestive tracks to spread their seeds. Adaptation would be necessary.

What is the grapevine all about? The scribes about vines and wine have been very creative in their poems and writings. Even up to this day you will see myths periodically regurgitated as eternal truths – even by people who should know better. Put simply – a grapevine is a plant just trying to survive and reproduce itself! That's all. When scribes write about wine being a natural product it is utter nonsense. The grapevine cares not about us and it is certainly not trying to make wine as an end product. To make wine humans must intervene in the process. Any winemaker can tell you that left to its own devices (or "nature") it will make acetic acid. Why in the world would it want to do that?

The vinifera grapevine is, as stated, a temperate zone plant. As such, that means it must have some mechanism for seeing that its babies do not drop on the ground, sprout and then be killed by the winter freeze. That would not be a good survival technique. The grapevine has many companions in the solution – at varying levels according to their normal area. Apples require that their seeds be below 32 degrees for at least thirty days for the greater part, to sprout. Blue Spruce needs below zero for forty-five days or more for sprouting. Manzanita seeds will lie for years until a fire sweeps through.

The grapevine, like the others but to a lesser degree, puts a hard shell around the seed. Then it follows one of two scenarios. The first scenario is when it senses that fall is coming it proceeds with "ripening" its fruit. When winds or whatever knock the berries onto the ground the wild yeast cells that have attached themselves to the cuten layer of the skin (the waxy coating) commence the fermentation process of the sugar- containing juice within the berry. The alcohol, in the presence of oxygen, proceeds to acetic acid. All of this takes time. The alcohol and acid begin to work on the shell of the seed weakening it and thinning it. This, too, takes time. The cooler the temperature the longer the process. The seed is now ready for winter – the temperature is hopefully below sprouting temperature and the seed can absorb some moisture from winter rains and snows. When the appropriate temperatures come back in the spring the seeds are able to grow, the sprout pushing through the weakened and swollen shell – prayerfully after spring frosts. Nothing is guaranteed which is why a large number of seeds are produced expecting that maybe a few will survive – or many in some years or none in another. The other scenario the vine uses is one for wider dispersion – birds and animals. The very sugar in the berries that we like is also appreciated by other animals and birds. The much wider dispersion is achieved by the droppings. In this scenario, however, the sugar is absorbed by the creature as a "reward" for dispersing the seeds. The acid attack upon the shell to begin the weakening process is accomplished in the digestive track of the creature. The same process is at work. Even raisined or dried berries on the vine in early winter, after fallen leaves make the berries visible to birds, will be taken – the reward is less but apparently enough to accomplish the vine's goal.

To accomplish this, the vine must have some mechanism or a combination of mechanisms to determine when fall is coming. That is, to determine when the time is right to complete the process and shut down for winter. The vine does have these mechanisms and they will be discussed later in this tome. Again – as in winemaking – human intervention is desirable to achieve our purposes within an understanding of those mechanisms.

The grapevine has two methods of propagating itself – three methods if we count the one where it charms us so much that we will stick chunks of wood in the ground but this is really an aspect of one of the two methods. The most obvious is the "vine" nature it possesses – as its canes grow across the ground and are in contact with the earth the vine will sprout roots at the nodes anchoring itself to the ground.

The vine can spread itself this way. If one cuts the cane between the nodes each rooted node will become an independent plant. There is one old method of filling in missing vines by this exact process. It is called "layering" and the method was to simply take a cane or sucker and extend it over to the desired location, dig a little trench, crack the cane and cover with dirt. The mother vine would support it until it was rooted. Then one simply cut it off from the mother. The other method the vine uses is, obviously, the seed method. The first of these is non-sexual reproduction and therefore should be true to the mother plant. "Should be" is not the same thing as "is". The seed method is sexual in nature and involves ovule and pollen. Many vinifera are self-fertile while there are grapevines having male and female plants. Vines grown from seeds are not true to the mother vine even if the source of pollen is also the mother plant. Seed-grown vines are the province of the "plant breeders" wherein they hybridize or cross certain pairs trying to enhance or decrease some characteristics in the progeny. It is a very exacting (exhausting), tedious and time-consuming process and best left to those who are enamored of it.

The evolvement of "clones" is a vine's method of adapting to new conditions and nature's experimentation. In a growing season each "bud" formed on a cane can become a new cane the next season. For some reason sometimes one of these buds will mutate – showing characteristics different from the mother vine in some small or large way. Some differences will be noted by humans and others escape our notice completely. Noticing differences was more common in the times and locations wherein one man or woman tended a small plot year after year even for generations. In today's corporate farming tending huge acreages with transient farm laborers and big machines differences in individual vines or grape characteristics go unnoticed.

The "genetic drift" that causes such a mutation is not clearly understood – at least by me and I doubt by anyone else. I do have some suspicions covered later in the Meador

Heat Pump Section. I suspect that someday a clever young person will find that one of the factors is chemical or temperature or both. Later I will discuss hormonal control systems I hypothesized at work inside a vinifera vine to cause the observed phenomena. The instructions to the growing vine cells are from hormones – plural.

These are different substances and different substances form at different rates and operate more rapid, or less, according to temperatures – I suspect. To change a bud's DNA there must be some error or difference in the controlling or directing mechanism or an effect of light quanta. It may, of course, be nutritionally affected but I think the temperature being experienced has a bearing. Of course, then the next question becomes why don't all buds formed at that time mutate? I just don't know. I do know that the subject buds decide the year of formation whether or not they will contain fruit-cluster initials and also the quantity of said initials. I also think that this is nutritionally caused but is also temperature affected. This I will cover later in the appropriate section.

You can observe a lot by watchin'
—Yogi Berra

Section 3B
ROOTS

Shortly after arriving here in Monterey in 1972 I was taken on a tour around and through a vineyard on the eastside of the Salinas Valley. The vineyard rows ran roughly east and west, perpendicular to the wind. As we circumnavigated the property I noticed that off-property to the middle south side was a gully. I also noticed that on the middle north side off-property there was a gully. I asked my manager guide "Doesn't it rain around here?" He asked why. I pointed out the gullies. He responded that that was unimportant, that they had just used belly-scrapers to level the land – filling the gully. I just said "Oh".

In the winter of 1973 Mother Nature re-emphasized her wishes concerning gullies! There were many different varieties planted there all on their own roots and the rows perpendicular to the water flow. Thus, I had the opportunity to study the rooting characteristics of various varietals hanging as high as sixteen feet in the air still attached to their wires and the soil hydraulically removed – not cut through by a backhoe. I took a couple of days walking the new gully making sketches in a little notebook and often stopping, sitting and just contemplating what I was seeing.

By this time I was very interested in roots because I was being told by "experts" and reading that one mustn't plant too close because the vines would compete and decline. Right from my start I was questioning the conventional spacing dictates. The American assertions simply didn't jive with the European experiences and writings. I needed to understand roots.

Let me again digress into apples (I'm sure you are going to get sick and tired of apples). The root pattern of apples is generally a *mat* of feeder roots solid perhaps, say, down to eighteen inches from the surface. They occupy most of the soil mass with fine feeders. Further, they then have main root branches going down much deeper for stability and also for water and mineral sources and resistance to droughts, if necessary, as a survival mechanism.

They also have another very interesting feature. Apples have "*mutually exclusive*" root systems. They actually exude chemicals that reject other apple roots! Upon contact with the exudate feeder roots will turn away. It is actually possible to "pot-bind" apples trees if planted *extremely* close in the field. For this reason the Dutch, in their meadow apple approach would tear out the field every so many years and replant. The apple tree generates a generous amount of root material and occupies a high portion of the soil.

My observations of the above mentioned grapevine roots revealed to me a completely different pattern. Those vines (across varietals) displayed a very small amount of root material. There were only three to five main branches heading downward. There were some side branches heading off sideways as if they had found some layer of soil of interest. The feeder roots – such as were there – were intact and very short hair-like devices. These were the post-harvest feeders. Remember, this was winter and the spring bloom of feeder roots hadn't occurred – a fact I learned later on.

Several characteristics of this study emerged. Grapevines are very poor root material generators and they do not occupy a large portion of the soil area but they are exceptionally efficient extractors. Their nature is to go down while exploring side layers. Another factor concerning grapevine roots came to me through French writings – namely: they are a *"mutually inclusive"* rooting system. That is, they do not reject other vine roots from their territory and, in fact, can intertwine like Medusa's hair! This is a very important realization.

The standard vine spacing for California, as touted by the schools and experts, was twelve feet between rows and seven feet between vines within the row which gives 518 vines per acre. Most vineyards installed in that time frame were planted thusly and you can still see a few dinosaurs in that format today. There were a few people that planted in the slightly different format of ten feet between rows and eight feet between the vines in the row – which gives 544 vines per acre. Essentially no difference. There was a French writing in the sixties – where they studied the California system and concluded that they couldn't achieve their desired quality of wine from it. Their asserted minimum number of vines per acre was about 2000. Now – keep in mind just because the French assert something doesn't make it any more factual than if a German or American asserts something. Their conditions, tests and rules were, and are, completely different than ours. But they had been growing grapes and making wine a lot longer than us and I think it is prudent to access the European mind whenever we can and then evaluate the thinking and observations.

A significant part of my time then was engaging in what I called "Myth Analysis". Nobody gets up in the morning and says "Today I will create a myth". So – where do they come from? Many seem to come from a series of limitations and ignorance. For example: a small wine grower grows a wine that the King fancies. The King sends a scribe to the grower to find out the hows and whys about the wine. The scribe arrives and asks his questions. The winegrower really doesn't know why his is different or has some technique he wishes not to expose. He does not wish to appear unknowing so he gives some answer maybe even a true one. The scribe is not a winegrower and, like all scribes, is limited by space and the understanding level of his audience (here, the King). He simplifies what he thinks he hears and a myth is born. My efforts were to ferret out the germ of fact or truth that is hidden by the mists of antiquity – that gem that was the basis. Then we had to evaluate its relevance to today.

Now – from where did this wide American spacing of vines come? Surely our forefathers knew about European spacings. Hell – they came from there! What were the factors at work to cause the change in approach? Well, the fact is that vine spacing varies all over the place in Europe as a function of local conditions. With our fascination with France - and specifically with Bordeaux and Burgundy historically – we have either ignored or not even been aware, until recently, of other areas.

The ability in modern times to push a button and make rain appear is a marvelous thing. But that is so recent as to have had no bearing at all on historical designs. Natural rainfall was the critical factor. And that had to combine with soil type – that is, the ability of a given soil to hold the water until the next application. In many of the European areas rainfall may come in two week intervals or so. Of course it varies all over the place but that is a pretty good working number. Locally, the growers will know their patterns and have matched soils to grapes and spacings that will work for them. There are soils nearby that they will tell you are no good for grapes. They will be rocky and sandy and hold no water or they will be very shallow with a hard layer underneath that cannot absorb water. The gist of it in Europe is that most areas receive summer rains to some degree or another.

In general, California does not – at least any that can be counted on. Thus, in California, like the plains of La Mancha in Spain, a different approach had to be taken – well, the same approach but with a different appearing result.

One must view "soil" as a water reservoir that just happens to be filled with dirt particles. Those particles and their organization will determine just how much water they can hold and yield to a plant. The old-timers of the mid and late 1800's recognized California's quirks and planted far apart. They hand irrigated for baby vines for a few years and then the vines were on their own – look at the spacing of the few remaining one hundred year old vineyards – free-standing vines widely spaced. This was all trial and error and "farmer guessing" by people much closer to the land than we are today.

As grapes moved into the Central Valley and flood irrigation became possible we saw trellising become the norm though there are still some free-standing vineyards in the Lodi area and elsewhere today.

But why the twelve foot spacing between rows? When it started I suspect that width of horse-drawn wagons had a lot to do with it but in later times tractor size certainly had an effect. Simply put – in America viticulture was such a trivial industry that there wasn't enough demand to warrant industrialists to tool up. Our industry had to make-do with tractors built for wheat, corn, sorghum, etc. Equipment was made (and is) by small shop producers in the Central Valley though this is changing a bit.

In recommending the 12 x 7 spacing the universities spoke of most efficient with least cost on fewest vines per acre. With the big vines of the hot Central Valley forming the basis of the studies, the standard was set. Keep in mind that this system was designed at a time when land was cheap and plentiful, water was cheap and plentiful and labor was cheap, plentiful and docile. Not one of those things was true in 1970 and subsequent.

In 1974, with overhead sprinklers in our hands, this spacing made no sense to me at all – other than equipment dictates which could be changed. However the vineyard companies had made their projections of costs based upon that spacing concept and part of the spiel to investors was that everything was in accord with U.C. Davis. Arguing for a different spacing would go nowhere as the initial cost side would go up substantially. So – in 1974 – on the property now called Ventana I planted two rows of each variety on the backhalf out of sight at twelve (12) by three and a half (3 ½) feet apart. This in – row spacing in our area made sense to me but we needed years to observe. These vines were own-rooted.

I will not comment a great deal on rootstock here as most of my experience is own-root. Monterey, in general, was planted own-root at the time. The general commentary at the time was that there was no Phylloxera in Monterey, heat-treated vines were only available as potted own-rooted vines and *if* we ever got Phylloxera there would be time then to re-plant with rootstocks because there would be hard wood for grafting. My, how everyone forgot all that later on when it arrived! The Ventana does not have Phylloxera and is mostly own-rooted. However, some rootstocks are used in certain situations. For example, Merlot does not like rocks and sand so it needs rootstock if grown on that type of soil – it likes some clay in the soil.

Much later on when the new viticulture was clear to me we adjusted the spacings in the row according to variety and soil condition. For example, Syrah on its own roots is a vigorous variety. We would plant from 3 ½ to 4 ½ or 5 according to the soil. Up over a weak rock bar the vines were closer. Over a richer low spot they might be 5 feet apart. I will address this again under the trellising category.

I did look at rootstocks early on and concluded that there simply wasn't enough known about them in my conditions. I also had some reservations about the printed information. I planted 17 different rootstocks interspersed with Chardonnay vines for comparative purposes. I just observed them until about 2000 or 2001 when I tore them out. If Phylloxera appeared I wanted to know which I would use.

One of those "reservations" came from my suspicions about Davis' strong recommendation of the AXR1 rootstock. Their position was not compatible with my studies of the matter which also created a gap in credibility on other subjects within my mind! The series of events is what I like to call the "Phylloxera Fiasco".

In the later-seventies there was a WITS conference (Wine Industry Technical Seminar) at which a U.C. Davis speaker was highly touting the AXR 1 rootstock for new or replacement plantings. I knew that many people had been using it (elsewhere in California – not Monterey) but I was aghast at the praise being heaped upon it. I rose, condemned the AXR as a disaster in the making and requested podium time to explain why. It was denied. I continued during breaks, and afterwards, condemning Davis' recommendation to small groups and individuals. I then gave up – who was I to be listened to? Further, it was of no real import to me or to my region to push ahead. If it did come up in Monterey I thought I could register some facts at the Monterey Grape Growers meetings.

Let's look at this rascal that caused (via Davis) so much apparent damage. First of all, its original name is GANZIN 1 – a cross developed by Victor Ganzin working at Montpelier (if memory serves) in 1879. At the time of its release Victor Ganzin clearly warns (like – "do not use in defense of Phylloxera" – pretty clear I think) against its use in defense of Phylloxera as it has vinifera in its parentage.

At Davis it was included in a comparative trial with many rootstocks. The study concluded that AXR 1 (Aramon X Rupestris) was the most desirable as its high vigor produced the biggest *YIELD*. Further, *even though this rootstock supported tiny colonies of Phylloxera with its vigor it simply outgrew them and their effects.* That report existed then – and still does if not expurgated in light of later events and somewhere in a hundred boxes I probably still have a copy. It was known at the time of recommendation that this rootstock was not completely resistant to Phylloxera!

As an academic exercise in recommending a plant one would expect some diligence in researching said plant and its performance elsewhere in the world. One would certainly think that, wouldn't one? In 1902 it was reported that

Phylloxera had seriously attacked Ganzin 1 in Europe and, though its great vigor delayed death, all such vines needed to be uprooted.

In 1980 and 1985 Professor Pierre Galet of Montpelier visited California and when made aware of Davis' promotion and touting of this plant strongly warned against its use and referenced the 1902 events. He was ignored. By the time he visited again in 1990 the signs of Phylloxera attacking AXR rootstock could be seen in Napa and elsewhere.

In the nineties the scientists "discovered" development of a "Biotype B" form of the bug that liked AXR and, subsequently, even a "Biotype C". They were "surprised". Of course, this mutated form is tendered as the "cause" and thus the earlier recommendation wasn't really at fault. It was just Mother Nature throwing curve-balls. Really? Well, the young scientists are not, I suspect, even aware that Phylloxera were living on those vines – their elders probably haven't told them. I also alienated, or just angered, some quarters when I responded to the excitement about these "discoveries" by stating "you are surprised about the emergence of a Biotype B? I imagine so. The concept behind its development is very new – it wasn't even published until 1859. Try reading Darwin's Origin of Species wherein he discusses the probability of mutation in response to a potential food source as a function of density of population. A few AXR vines no sweat but you guys pushed AXR. A billion vines is a whole different matter – there is your density." I suspect that it wasn't even a new emergence or mutation. I think those little rascals in the test block told the story. A certain small percentage of the population was already amenable to the vine. I am of the opinion that the "Biotype B and C" are not really a mutation in response to the AXR food source at all. They were there all along as population variants. Otherwise, how does one explain the rapid changes by 1902 and the California exercise? *Widespread* spontaneous mutation and devastation within 20 years? I think not! I think the tendering of the biotype story is a CYA spin.

We are all aware of the great uprooting of vineyards almost overnight in the 90's, all the press stories and the loud roar of misery by growers and wineries. Phylloxera is a funny little bug and has created problems in the past – vineyard tearouts. Nothing new. Problems from Phylloxera are more like AIDS – not like a heart attack. Why the massive rush to tear out vineyards immediately? Why all the loud crying? Vineyards are often torn out or grafted over losing crop years. This is normal to the business and vineyards can be re-done sequentially over time.

We should consider a few apparently desperate facts to perhaps find some answers.

a) The new viticulture was beginning to be accepted and awareness was beginning to penetrate that it gave higher yields at a higher quality and a lower cost per unit of production. If one farms in the old way one will have a lower quality product at higher cost – and Rolaids won't cure that.

b) Federal tax laws say, simplified, that if you change your vineyard because you want to you can't write off the remaining capitalization on your earlier planting nor can you put your new planting on an accelerated depreciation schedule – both are "forms" of financing or money. But if you do so in response to a bona fide disease you can!

c) California law didn't even allow that – it didn't track the Federal format. If we yell and scream as a group loud enough maybe we could get that changed. *It was so changed.*

d) Were the AXR vineyards replanted on other rootstocks in the old format? Look around. They were replanted in the new system. There were vineyards that had never seen an AXR that all of a sudden were declared AXR rooted! AXR became the villain for every problem and excuse to the bankers for a tearout. Poor old AXR – a disreputable, hated mal-performer that should never have been on-stage in the first place – and all because homework wasn't done by the academics or others. Incredible!

e) The chorus of voices was full of honorable people who had simply been misguided and needed assistance as the American Banker had turned on them long before. The chorus was being quietly conducted – and conducted well – by those who understood the situation as a needed raid on the public purse. And that was okay in my mind – so what if the large conglomerates received benefit? The small independents *needed* it more in order to survive and America would receive better wines from better grapes. When we consider all of the above it is possible we may conclude that the AXR fiasco provided the horse to bear the burden of the massive vineyard design change and the revolution in viticulture.

In 1980, or thereabout, the academics (Kliewer, Fideo, and Kazimatis) reported that in one of their studies that 9 foot rows out-yielded 12 foot rows from .9 to 1.6 tons to the acre but that was not their focus – they just noted it in passing. In conjunction therewith they also mentioned that while there was more cane growth *per vine* on the 12 foot spacing there was more *per acre* on the 9 foot format. Their main focus was upon "T" top training and "Wye" shaping upon the old standard existing system. They reported again on the better yield of AXR #1 over St. George and 99R, terminology which any practitioner of the time would take as favorable for AXR#1 use. They also referred to AXR#1 as "the standard for comparison". There was simply no attempt to re-think and re-design the entire system. The academics then, and for the next decade, were tinkering with a "Model A" not even realizing that a "Corvette" had passed them some time before! In 1985, when they "tested" shearing they did it within the old system not the new system designed to receive it. Of course, they cut off half the leaves – not only that, the leaves they cut off were the mature CH producing ones! And from that they concluded that "shearing" delayed maturity – not the wrong removal of CH producers. Tinker, tinker, tinker. Why tinker with these little static projects? Because there was nothing in their training or culture that would lead them into dynamic analysis or a re-working of their academic predecessors. To do so would also have led to their academic ostracism and professional demise.

Many Americans believe the writers' myth that all western European vineyards are on American roots and that you must have rootstocks to protect against Phylloxera. Well – that simply isn't the case. Actually there are quite a few vineyards in Europe (and here) that are on their own vinifera roots. They have one thing in common – they are on sandy ground with less than 7% clay material. At 5% to 7% Phylloxera have difficulties. Over 7% you should have rootstocks. There are other vineyards on heavier soils that, so far, have been isolated from the plague. However, there is another problem - nematodes. Unfortunately, nematodes love sandy ground and can be devastating if ignored. Sadly, our vine breeders have not yet been able to give us a rootstock that is immune, or even strongly resistant, to both Phylloxera and all nematodes. Even Freedom rootstock is susceptible to some nematodes not to mention its weakness in picking up zinc.

The understanding of how a grapevine roots and occupies the soil was an important part of my closer vine spacing concepts. The fact that "mutually inclusive" rooting plants that do not generate large masses of material in the soil allotted allows the closer spacings. They may not be big rooters but they are very efficient at what they do. Further, it clarifies for the mind the capability of the vine to perform under a wide variety of spacings under different soil/climate conditions. It focuses the importance of those conditions as the primary controlling factors for consideration.

The root tips are the important generating points for a family of hormones called "cytokinins". These compounds control and/or contribute to shoot growth, fruit set and fruit-bud differentiation (possibly). The quantity produced is a function of available nutrients and, I suspect minutely, soil temperature.

*That is the essence of science: ask an imper-
tinent question and you are on the way to a
pertinent answer.*
—Jacob Bronowski

Section 3C
PHOTOSYNTHESIS

For all practical purposes, photosynthesis is the source of relevant life as we know it (I'm ignoring sea-floor vents). The
green tissues of the vine manufacture the food the plants
need – carbohydrates – by photosynthesis. The process is a
chemical one and its rate of activity is a function of temperature. Someone once told me that chemical processes
double in rate for every 10 degrees centigrade increase in
temperature. The process also involves taking air into the
leaves through "stomati", or tiny holes in the leaves, using
the carbon portion and exhaling oxygen. When the stomati
close the gas exchange reduces or terminates even if the sun
is shining.

The vine has a method of sensing adverse conditions
that could dehydrate the vine beyond its ability to replen-
ish water from the ground. When it senses those conditions
it closes its stomati to protect itself from water loss. It does
this by comparing the moisture pressure within the aperture
to the immediate surrounding air next to it – a layer of air
called "the boundary layer". In the calm that differential is
a gradual scale not bothersome to the vine. Winds will strip
that boundary layer away creating a large differential and
the resulting closure. This was known to me in 1974 – in fact,
it's right out of plant physiology text books from Cal Poly. It

was why Sudax was used. Later, its one reason growth tubes work.

The afore mentioned tremendous increase in apple yields and quality was not from some magical fertilizer but rather the new orchard designs gave a huge increase in *leaf surface exposed* to light! The critical factor is exposure. The next factor is *quantity* of *exposed* leaves per acre. If you want to increase sugar beets or apples or wheat figure out a way to increase the per acre yield of photosynthesis substrate – that is, exposed green surface per unit area. It all comes down to grams of fruit per hundred square centimeters of *exposed* leaf surface!

The cold winds and their effects upon the vine had other little surprises for us. Terrell and I saw bloom take place in excess of thirty days across a vine and as much as twenty days across a cluster! I am speaking of primary crop not second. That meant that for a crop that takes 120 days to average some degree brix large portions were unripe. Sound a bit like "veggies"? Unripe flavors? The machines so in vogue took them all. In 1977 and subsequent – until we had the new system in place or using other techniques – I solved that problem by using the old Common Delicious apple technique of selective hand picking. We would often make three passes across a block and no veggies appeared in the wine. The first sets of commercial gold medals at the judgings were in 1978 from the 1977 crop – the first commercial crop on Ventana. Those grapes were harvested in that fashion.

But that, in my mind, was just remedial, correcting a problem short-term that should be handled by better vineyard design to achieve better uniformity. That is, a smaller deviation from the mean brix. The goal had to be to reduce the full bloom period to ten days or less! Our industry could not survive with the hand-selection multiple pass nonsense. There had to be a better mousetrap somewhere. I had to find it.

The wind problem was well-noted in my mind and would have to be addressed in any subsequent design. I will cover this solution in the section on trellis.

In those years Amand Kazimatis – an extension agent from Fresno State – would come to Monterey from time to time and hold sessions trying to teach all of us wannabe grape growers about grapes. He was a very charming and knowledgeable man and made a significant contribution to Monterey. He was from Fresno and well versed in the extant viticulture of the time and place. At each gathering I would ask some off-the-wall question then reference apples as the reason for my question. After many of these, one day Kaz – in frustration with me – said "Dammit Doug – vines are vines and apples are trees and vines and trees are different. You must start understanding that!" I said "Jeez Kaz – they both have green leaves. Maybe we should look at what those folks are doing". We dropped it and Kaz and I went on as friends. In those days the wine industry didn't look to others. All that was worth knowing was here in California – eastern grape growing was dead but nobody had told them yet and, besides, they only grew those "foxy" tasting grapes the wine from which no civilized person could swallow. They were non-entities and if they did have some knowledge it was useless in California. That was the general attitude along with the idea that there was nothing to learn from other crops because wine grapes were so unique. It was incredible, I thought. My family had taught me to be an equal-opportunity thief – steal a good idea even if from the devil! The Navy had taught me to question all authority. In 1973 and 1974 as I looked at trellis and vineyard designs throughout the world I kept coming across the name Nelson Shaules at the Geneva Experimental Station – a branch of Cornell University – in Geneva, New York. Shaules was the man who developed the Geneva Double Curtain trellis and that had radically changed the labrusca (concords and niagaras, etc) industry there. The system had not only improved quality but doubled and tripled yields. Now there is a strange thought – better quality at higher yields. Most curious. Probably one of those Eastern ideas, good for jelly production.

Shaules was deeply into the study of light penetration and leaf exposure of grape vines. In 1975 I decided to go to Geneva, meet the man and hear from him what this was all about. I also intended to pick up anything he had written. Geneva is also a major apple research facility and I wanted to gather all their writings that I hadn't read. Folks – it wasn't the day of the internet, this is how we did it then (or by mail). I did so in the early summer. When I reached Geneva I found Professor Shaules' office on the second floor, in a corner, in the Pomology Department building (Pomology means apples) surrounded by all these folks doing research on dwarf apples and light! It was no surprise to me – now I understood. My impression was that he was the only professor working on grapes.

The stage of his work then was that he had a method of converting pictures taken inside a grapevine looking out to light penetration analysis. He had a couple of grad students crawling inside vines with 35 mm cameras. In November, 1974 Professor Shaules had presented a monograph to the American Horticulture Society and he gave one of those two grad students sub-authorship attribution. That student was one Richard Smart of Australia. Much later, now Dr. Smart, he went to work for the Ministry of Agriculture and Fisheries in New Zealand conducting grape and light work. Perhaps a decade or more later he published a book called "Sunlight into Wine" – a *must* read for any winegrower. More than just "read" – memorized might be a better move. Throughout the intervening time period he presented his work in various writings and at conferences such as the first cold climate Symposium held in Oregon in 1984 and at U.C. Davis in 1986. He has made major contributions to the wine world reflecting and expanding upon Professor Shaules' seminal work and teaching.

Yet, as excellent a rendition as it is, it is only a small part of the picture and an easily discernable one at that – at first glance. It is a beginning only. The inter-relationship of light (and its associated heat) with the hormonal and nutritional matrix has not yet been addressed.

By the time I arrived in Monterey in 1972 the Mirassou Mission Ranch and the Paul Masson Pinnacles Ranch had mature vines. I spent considerable time just looking at them with their 12 foot spacings and contemplating the wasted space between rows. It seemed to me that a lot of sunlight was hitting ground – not green leaves. In the winter months I would walk the rows sketching in my trusty notebook what I was seeing. I find that sketching focuses my mind and forces it to become cognitive of the thing. I find that many people look but don't see. "Looking", to me, is a mechanical thing while "seeing" implies cognition or understanding. Consider our response when we understand something – "Oh, I see".

Monterey had its wind problems that drove the canes to the south piling them up one upon the other in layers. I was told by academic "viticulturists" that the second and third layers were still functioning well – "probably" as much as 70 – 80% of the outer leaves. That didn't seem probable to me. At different times of the year in 1973, 1974, 1975 and 1976 I would go to Napa, Sonoma and the Central Valley just to look at grapevines and try to "see" what their structural characteristics at work were. In those situations without the daily hurricanes the appearance of the vines was considerably different. They were more like long narrow haystacks with the canes going up over a wire then swooping downward often growing across the row center! The fruit grew in a dark tunnel down inside the vine. From mid summer on, the basal leaves had senesced from shading out. In 1973, and for a few years, many Napa and Sonoma vineyards were still free-standing vines though trellising of the vineyards was proceeding rapidly. A fellow by the name of Cesar Chavez was scaring a lot of growers (rightfully so) and there was a big push to make the vineyards compatible with the new harvesting machines.

Then, though, one could still have lots of opportunity to do comparisons in same soil situations. One thing that struck me was though at first glance it appeared that the same thing was going on in both systems it really wasn't. In the California high-wire system the vines grew upward and over a wire strung about a foot above the *cordon* and *then* over and down. It was in that foot or so area that the fruit was concentrated and shaded and later the leaves senesced. In the free-standing vine the curvature downward begins almost immediately and diffused light is not only on the clusters but good light is on the basal buds (SEE PLATE 9). Subsequently, the ones I have seen in Spain and France behave the same way. I will address these matters a bit later in the hormonal systems and trellising sections.

Row orientation is an important matter particularly as the rows get closer and closer. Rows in-line with the azimuth of the sun are receiving light directly down on the tops and not penetrating the vine from all angles. The fruit is essentially shaded as are basal leaves. Rows perpendicular to the azimuth receive light at varying angles throughout the day. Unfortunately there are other considerations one must take into account such as winds and their direction and, sometimes, the shape of the land parcel if smaller. Again, apples were my guide and in 1973 when I laid out the format for the irrigation system I simply stuck a pole in the ground and then marked where the shadow hit throughout the day. That gave me the sun's azimuth in my location. Pretty high tech! It works. It was also the way I had been taught by my dad when he laid out his orchard. I laid out the Ventana nearly perpendicular to the wind and about 20 to 30 degrees magnetic angled off the azimuth which is a nice tradeoff. This pole business will come up again in the trellis section when I discuss shade patterns. The wind aspect will also be discussed further.

Section 3D
SPACING AND ORIENTATIONS

As I mentioned elsewhere row orientation becomes a matter of trade-offs in the judgment of the grower. The wider the row spacing the more amenable the selected system is to variations in terrain without encountering shading out from one row on the next. I will confine myself to the flat as practitioners will vary the concepts according to need. I am presupposing a row perpendicular to the sun's azimuth in the sixty days before harvest.

Consider an acre of land upon which we put one row. That will give a production of photosynthesis substrate of x. As we then add row after row and plot on a graph the substrate production we will see a climbing line increasing arithmetically. SEE DIAGRAM C NEXT PAGE.

At some point as the rows get closer and closer one row will *begin* shading out some of the lower leaves on the adjacent row. However, the gain from the new wall of leaves is greater than the loss from shading. Thus the graph line will start to change from a straight line to a curved one. No longer is the gain equal to one row additional. Here, remember, we are strictly speaking of substrate production per unit area and not *now* considering wine quality as a result of light on the fruit. Continuing the mind game, at some point as we get closer the gain from the wall of leaves will equal

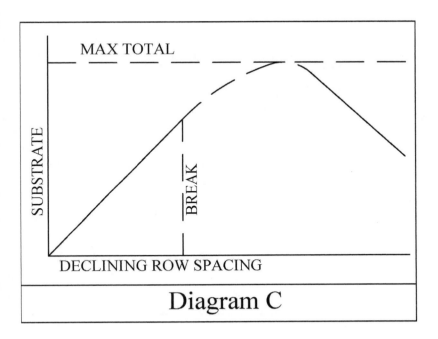

Diagram C

the loss from shading. That spacing will be the maximum value of quantity of photosynthesis substrate per unit area for that row design. From then on closer rows will yield reduced amounts and the curve will be in a downward direction. This pattern is true of all crops, I believe, and sets the limits. Unfortunately there are a variety of other limiting factors such as the width of necessary and available types of tractors and equipment, skills of people, effects of light on fruit (in our case), etc. etc. that dictate that we are not able to work at that maximum point.

Earlier I mention setting the pole in the ground and marking the shade pattern. If the pole is of the height of one's proposed foliage height at maturity this process will show one the length of one's shade pattern. The application of this knowledge is variable and arguable. The writings I have read concerning the spacing almost uniformly conclude a 1:1 ratio. That is, if the wall is six feet tall the row spacing should be six feet. The one glaring exception was some French writings stating 1:.8. I think that this can be explained though I've not seen any French explanations – if any exist.

Usable sunlight is measured with a term called "Langleys". In the early and late hours of the day the light travels a very long distance through the earth's atmosphere which filters out much of its plant-growing usefulness. In addition, the morning time is usually fairly cool so the usefulness of what usable light does get through is diminished by a reduced *rate* of photosynthesis activity. But there are further aspects to consider.

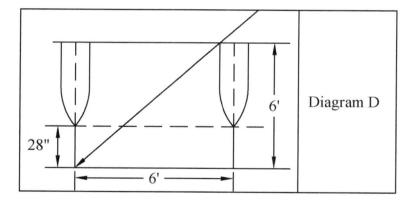

Diagram D

Let's assume, for exercise, a six foot height and a cordon height of twenty-eight inches. The 1:1 assumes a usable light pattern as shown in Diagram D.

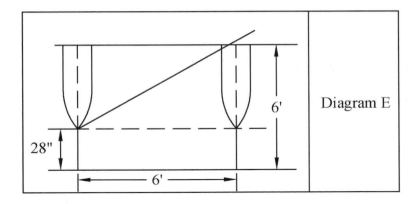

Diagram E

Consider, though, that there are no leaves below the cordon wire so it does not matter if the trunk area is shaded. With this recognition the pattern could be closer.

Further, if one intends to *fully* strip leaves to 10 or 12 inches (which I do *not* do and do *not* recommend) as I observe many practitioners do then that light angle would be even flatter allowing even closer rows before the bend in the curve occurs.

By the middle seventies I had concluded that six feet between the rows would be proper for Ventana and for others however at that time there was no equipment readily available to handle such a vineyard design. In 1978 I did install a few rows in that pattern to commence the testing of my theories. The equipment problem did put limitations on our layout but we adjusted. We planted a row down every other center so that we had three rows six feet apart and then a twelve foot space until the next row. This allowed us to get our equipment through. We put nozzles on our sprayer to point out so we not only sprayed each straddle row but one side of the in-between row.

When picking (by hand) we passed the grape buckets under a row to the gondolas in the twelve foot area. We pitchforked pruning brush over a row into the wide area for brushchopping. It was not until 1982 that we installed area planting in this fashion on seven acres of Riesling and seven acres of Chardonnay. Those vines exist today. In fact, the cordon heights are higher than 28 inches – but lower than the California high wire system as I had not yet determined that the 28 inch was correct. We later tried 24 inch but that was too low. It was ergonomically incorrect for workers at harvest but more importantly at pruning time. The 28 inch position allows people to have straight backs when the vines are mature for all work except suckering.

The six foot row spacing is a little tight and I regularly spoke publicly about recommending seven foot spacing for corporate farms where labor and management skill might not be adequate. Today, with stable workforces, that probably isn't the case. Back then migrant labor was the norm. Management was essentially young, English speaking and unskilled in teaching. Today grapegrowing has provided a nice offset to regular row cropping so our farm labor is now heavily residential.

In-row spacing is another subject. Keep in mind that the prevalent spacing was seven feet between vines and twelve feet between rows. Once the workability of the six feet row spacing was determined it was an easy conversion from twelve foot – simply plant a row down the middle and intercept all that lovely sunlight! Necessity is often the mother of invention, as they say. That was a no-brainer even if people accused me of that – no brain. Actually, they didn't accuse me of that. Rather, they often and loudly said I was crazy and it would never work. When I spoke about the concepts at our grapegrowers meetings I would be looked at as if I were from another planet. By the way, Peter Mirassou and I started the Monterey Grapegrowers Association. Peter was the first president and I the second.

Proper in-row spacing is more difficult to achieve and requires very good knowledge of the soil, vines and rootstock. I don't think it can be achieved perfectly – "perfect" keeps hovering out there like one of Plato's "ideals". There are simply too many variables. But we can, and should, make every effort to strive for it.

Let's look at the old seven foot vine spacing. As I stated earlier I would study the in-row mature vines of Paul Masson, Mirassou and others. The methodology then (General Viticulture, Winkler et al) was to grow a vine up a stake – each vine to have its own stake. At the end of the first growing season the vine was cut back to its base. The next year it was grown and tied up the stake. From their central valley warm climate experience when the vine was above the cordon wires it was headed off at or below the wire to force side shoots which would become the cordon-forming canes. That was to take three years. In Monterey it regularly took five years! The "heading off" severely set back the vine and a year was lost. Heading off and forcing side shoots to become the cordon created a "yoke" or Y shape at each vine. The stated purpose was to "balance" the vine distributing sap equally to both sides on runs of "3 ½ feet". It was recognized that runs of seven feet would lead to diminished growth the farther out one went. That is, too long a run would diminish the nutrient flow – just like a too long pipeline with sprinklers would deliver less water at reduced pressure the further out one went. This is an important part *not* spelled out. It was inherent in the writings but not recognized. This *part* must be kept in mind. It was important, in my mind, once I recognized it in understanding phenomena I was seeing in fields. It still is today. It will come up again in the pruning section and when trying to "perfect" the in-row spacing.

My observations of mature vineyards then – and still today – is that when the "yoked" form is used the distance from the opposing shoulders would often be 12 inches or more. I have seen some growers criss-cross these cordon shoots in an effort to reduce that gap. That approach does create a little congestion in that area if accurate dis-budding is not done. Further, here in Monterey there would often – no, usually – be as much as 12 inches or more from the end of one cordon arm and the end of the next one.

When we construct a trellis it is for one purpose and one purpose only – to hold up grape vines. That seems so simple but even today I see trellises holding up air! Trellis costs money – so does the land it is on. We irrigate, fertilize and spray the whole area. To not utilize every running inch of that cordon wire is to operate below *potential factory capacity*! Have no doubts – that is exactly what a vineyard is – a grape factory. That cordon wire is our production line and to the extent we do not utilize it optimally we underproduce. If it is because of poor design for the conditions the design must be corrected. In this case consider that a 12 inch gap above the yoke equals 14.3 % of the trellis footage allocated to that vine not being utilized! Consider further that an additional 12 inches per vine (6 inches at each cordon end allocated per vine) is also 14.3 %! That is 28.6 % of the expensive trellis not being utilized!

Recognizing this startling implication of gaps and my observation that, for example, it appeared that mature Chardonnay own-rooted on my soil and climate could support about 5 feet or so of *equivalent* cordon a design was developed to correct the economic deficiencies. It is immediately obvious that one wishes a production design that is *continuous* from end post to end post with no dis-continuities along the run to achieve maximum potential yield. As an aside – as I was taking pictures in February, 2008, in various regions of the state I was astounded at the number of vineyards, both young and older, that have completely missed this point!

What were the gains of such an approach, if achievable? In those days Napa County Chardonnay ran about 4 tons per acre and Monterey ran about 3 ½. Many (most) winemakers thought that 2 ½ was the proper quality number! I'll discuss this in the Yields Section. For now, consider that such a design would have two lines of fruit/vines where before there was one. That alone should double the yield per acre to 7 or 8 by arithmetic. Now – if one was able to increase the efficiency of the in-row situation that should increase the total production by another 20 – 25 %! In a Region I! 8.75

theoretical tons per acre (3 1/2x2x125%) It was "Heresy" of the first order!

Once the underlying principals were recognized and understood in total the answers seem to leap up screaming to be noticed. The presence of the overhead sprinklers canceled out the soil-moisture reservoir historical limitations for consideration. It was replenishable at any time. Likewise, nutrition could be introduced if needed.

In the mid-seventies I drafted out a little equation to help me think:

$$C+S+P+FP+WP=W$$

Wherein:

C = Climate

S = Soil

P = Plant

FP = Farming Procedures

WP = Winemaking procedures

W = Wine

This summation of our work lays out the farming factors of our business to the eye even though we all subconsciously know them.

Consciously listing the factors, and the studying of each thoroughly, aids the identification of mis-steps, prejudices, faulty thinking and irrelevancies.

As I studied the practices of various parts of the world some things became salient. *Within* regions or districts in Europe there was little or no variation in farming procedures even to the extent of codifying them in France and Germany. People did things the same way, deviations were frowned upon and social pressure was brought to bear on miscreants. Change was not prized. Even here, as much as we like to think that we are "free" to do as we wish, there is a distinct uniformity to procedures and equipment. Winemaking procedures are also the same within regions. The same factors noted above apply. Again, even here, that is true. The schools all teach the same thing and corporate mentality limits explorations. Thankfully we do have "artistes" or "loose cannons" in our tiny wineries exploring limits and/or new approaches. More of this is happening in Europe now. However, they are self-funded, usually, and broke, usually, but they are having a ball and loving life.

In the above equation CLIMATE is a *variable* – everywhere. Soil is essentially a constant except for nutritional additives that are usually fleeting. The structure is fixed once the vineyard is installed. Irrigation can cause a slow salt build-up depending upon water quality and practices but in the reasonable run it is a constant.

PLANT is a constant once the vineyard is planted. Of course, it can be changed but that sets up a whole new scenario.

FARMING PROCEDURES are a constant for macro-analysis. Upon close observations there are really no big differences between operations within a region.

WINEMAKING PROCEDURES, in general, were historically constant in Europe within regions.

Given the above, a simple mathematical analysis of the equation by someone with no knowledge what-so-ever of wine would conclude that with *one variable only* there would be better wines some years and poorer ones in others as the annual weather varied. When we bring that conclusion to the wine cognoscenti their response is "Of course, fool. That is a well-known aspect of the wine world", and they would be right.

The thing that occurred to me was what if we could change Farming Procedures making it a short-term variable from a constant? What if, in fact, we could vary farming procedures enough to affect, totally or sufficiently partially, the climate variable? Wouldn't that be something? Actually, we already did that somewhat by frost protection but other that that I don't think we growers thought that way. We would "change" some scheduled activities to fight a pest or mildew but at that time that was it.

Consider one example of spraying under the old Davis system. With the haystack of canes and foliage coverage was a problem and diseases lurked down inside the vine waiting for the right conditions. During my early years in this industry I thought "wow, these organisms are something else if we are diligent in killing them and then climate conditions are right and that one last bug can hit the whole state in a matter of days! No, that can't be. They must be there, lurking and waiting." They were. Go back and look at the extension notices every year and several times a year – they regularly told growers to be sure and get coverage, be sure to test and clean your nozzles, be sure to use plenty of water per acre – as much as 600 gallons per acre in some cases. Over and over – every year, year after year! The chemical company guys would come out and look and, for sure, the grower wouldn't have "coverage". The recommendation? More chemical and more water! The equipment guys were building bigger sprayers that took more horsepower and bigger tractors. In those days we didn't have the designer chemicals extant for some years now but if resistance develops the industry could revert. In those days when mildew or botrytis raged – it *raged*! It could devastate a crop. Monterey was "Botrytis Central" with our general conditions.

What if we could change a farming procedure to handle this problem? What if we could re-design the vineyard so the canes did not pile up in a haystack? What if we could have the canes out of the way and our target zone *open* to our sprays? Open? What? If that could be done then the implications were immense. If the zone was open one would have the natural precluders of disease working for us – air and light penetration. Apparently bad bugs don't really like dry, well-lit conditions as much as they like dark, moist areas to do their evil! Smaller amounts of water would be needed, smaller fans or pumps would be needed and that meant smaller tractors and that meant less original costs and less fuel usage and less soil compaction! Wow!

An example of changing farming procedures in response to a weather change is irrigation usage in a hot spell. The vines in Monterey are used to growing in cool conditions. Every so often we can have an attack of heat – 100 to 105 degrees! This can be damaging to our grapevines and for sure to the wine. I have seen at these times not only widespread tip burn and leaf burn but also vines completely defoliated except for the last few leaves on the canes – with the blue grapes hanging all alone at 18 or 19 degrees brix and cooked! This occurred several years ago (two years in a row) in a block in a vineyard nearby Ventana! I took pictures of what can happen. At Ventana we had no damage from these events. Why? When the heat came we would start about ten in the morning turning on every fourth sprinkler line. Then, after an hour we would turn on the next line and turn off the first continuing this turning on and off until about 4:00 o'clock in the afternoon so that the vines would dry before night. By measurement we hit a few times 20 and 21 degrees *reduction* in ambient air temp compared to the temp in adjacent vineyards! Usually it was less but effective. We not only protected our tender leaves but also protected the wine by not allowing the heat to metabolize *malic* acid or to burn or cook the grapes.

Another in the opposite direction was 1989 – the year of the "rots". Botrytis and other rots had attacked with a vengeance with the rains. It was quite the media story. When a local TV station got the story of what was happening up and down the coast they called me, which they often did, for the lowdown. When I explained the devastation that was occurring they expressed sorrow for me. I said "Whoa – not me. We have none". They were shocked and wanted to know why. I said "I am blow-drying the vines". They asked if they could come and film and they did. We were simply using our sprayers and sulphur dusters empty – just as blowers to blow the waters off the grapes. At that time I was selling some grapes to K-J. Their field person called me up and said "Well, I guess we're not getting any fruit from you this year". I said "Why not". She said "Everything is rotting everywhere". I said "Not on Ventana". She arrived the next morning, inspected the crop and we sold them the fruit again that year. Our other buyers were equally pleased.

The problem with the old Davis system was that it, in fact, enhanced the problems. What we needed was an entirely new paradigm of viticulture – one geared to cool climate based upon the physiology of the vinifera grapevine.

With modern understandings we also now can make WP a variable to some degree. In many places there are laws and regulations in place that limit but, in general, there are tools available to correct poor farming (or unlucky farming) particularly by the producers of industrial wine.

---

*Often, the less there is to justify a traditional
custom, the harder it is to get rid of it.*
—Mark Twain

"A new idea comes suddenly and in a rather intuitive way. But, intuition is nothing but the outcome of earlier intellectual experience"
—Albert Einstein

Section 3E
HORMONAL CONTROL SYSTEMS

In 1974 I was beginning to realize I was in a new world for me and I needed to approach this one just like I had my last. In flying I focused on knowing everything there was to know about my airplane – every rivet and wire and screw – largely because my life depended upon it. I didn't really trust luck although I do have to admit that luck certainly played its role at times. I bought, from various university bookstores, the first two years' textbooks on plant physiology. Jeez, working my way through those was difficult. Periodically I had to call up friends to explain stuff to me. Happily, now some thirty years later, I've forgotten most of it and those dog-eared books are long gone and far out of date. I simply learned it, sifted it, took what I needed and applied it. I still don't understand how one can get a viticulture degree with one course in plant physiology! After all – a plant is what it's all about.

Very basically a hormone is a *regulatory* substance produced in an organism and transported in tissue fluids such as blood or sap to stimulate specific cells or tissues into action. They are chemical messengers that act as catalysts for chemical changes at the cellular level that affect growth, development and energy. Hormonal systems within the vine control the vine's activities. I knew that the apple researchers had hypothesized the existence of certain hormonal control systems within the apple tree to explain their observations. Normally, I don't wander about contemplating hormones but then I had to in order to construct an operating theory that would elegantly (in the mathematical sense) explain every observation, wherever made, of the vinifera vine – or at least, sufficiently so for the purposes of a grower. The main thrust was not just to explain but consistently provide predictability.

Over the next three years I put in as much time as I could spare in what I called "river-bank time". I would go to the river, away from phones or questions from employees, and just sit and think through this inundation of information I was receiving and try to put order to it and make sense of it. There are many systems at work, I'm sure, but most of them aren't necessary for we growers to understand. Actually, I think nobody understands them yet.

One such system that may or may not have some implications for growers when it is understood, if ever it is, is the sap circulation system. There are several theories. One involves the transpiration loss of moisture by the leaves to the atmosphere creating a lower pressure in the leaf area "pulling" the moisture upwards from the roots which are "pulling" it from the soil – all a function of lower pressure pull. Another is capillary action. But capillary action is really a pull not a push.

Here I would like to introduce a concept I call the Meador Heat Pump (MHP) to address many issues.

Back in the seventies and eighties there were two explanations tendered for the circulation within a plant – both unsatisfactory because counter–examples existed. I haven't read plant physiology books since the early eighties but I doubt there has been improvement – at least to my knowledge. At the time I developed the MHP as a working hypothesis for explanation of certain contrary conduct by a grapevine as well as other plants.

One puzzling aspect of a grapevine is the appearance of a "pressure system" within the vine. I have implied this elsewhere by referring to "push" and distribution of energy – both stored and currently produced. But there were gaps not adequately explained by the extant theories. These "gaps" defied the conventional wisdom and required filling or complete revision such that we could understand more fully the activities of the vine.

The two aspects of circulation were, simplified, "pull" from transpiration losses through the leaves and capillary flow. Both of those are very slow processes.

We have all seen, during pruning season, certain different vine responses. In one, when the temperatures are very cold and the vine is in deep dormancy, we make the cut and the wound stays dry. In another, we make the cut and the wound may seep sap – perhaps not in the morning hours but in the afternoon or as soon or shortly thereafter making the cut. This seepage or weeping will run down the spur for a day or two before sealing the wound. Finally, we have all seen the vine "spurt" sap as the cut is made – sometimes as much as two inches out, indicating significant pressure.

At this time the leaf transpiration pull is certainly not a factor. So – in the second instance – how in the world would capillary action apply as there are no capillaries beyond the cut to create any "pull". In the third case capillary pull would not create pressure causing a "spurt".

In my opinion the grapevine –as are we humans – is a closed system and must remain closed in order to operate. When we are wounded defense mechanisms kick in to seal the opening as quickly as possible. The same with the grapevine. There is a portion of the sap that, with contact with air, changes into a gooey jelly-like substance that seals the weeping wound. In the situation of no bleeding the outermost cells "dry back" but the sealing takes place a bit further back. Remember the admonition to cut through the bud above the one you wish to keep so there is no drying back?

This closed system is very important in my mind and explains many phenomena in the field. I have immodestly given my name to the concept because I believe it has not been propounded nor elucidated elsewhere and it was for my own use – until now. If I am in error I deeply apologize for my ignorance of the fact. I have not read on plant physiology current thoughts for a great many years and this line of thought occurred in the mid-seventies.

The system, I believe, operates as follows. As a temperate plant, the native areas of the grapevine receive winter moisture while the plant is dormant. Dormancy is a requirement of the vine and we know that a certain amount of "cold-soaking" allows it to perform best. During this cold, wet period the vine absorbs moisture. The absorption action of the wood creates a "pull" aspect through the vascular tissues. We know that absorbtion by wood is a powerful force. The ancients, up to the near past in order to quarry stone, would drill holes and drive dry wooden pegs into the holes. They would then drip water onto the pegs which would absorb the water, expand and break the rock. In the case of the plant the winter absorption would saturate the tissues with COLD water. With the coming spring associated with the shifting sun (actually the Earth but what the hell) more Langleys of sun energy are directed at the vine. With the approaching spring there are wide variations in conditions. Some days are still cold, some are cold in the morning and warm for a few hours in the afternoon, some are modestly warm most of the day, and the next day can be like any of the above.

As the "quanta" energy packets are absorbed by the water molecules and increase their energy levels the electrons move to further out orbits and thus "expansion" in area. The higher energy state translates to heat. The effect of that warming is to increase the temperature of the vine and thus the liquid in the vine which is at its maximum concentration for the cold winter temperatures. As the liquid warms it *expands*. The warmer it gets the further it expands. *That heat – induced expansion creates pressure within the closed system.*

That pressure forces the young shoot to break through the sealed, protective shield of the bud.

This is the mechanism with which the vine determines that spring is arriving. It is not some sort of average over 50 degrees F (still used today for a factor in regional classification) but rather the temperature differential mechanism that returns the vine to an awakened status.

Most growers are pleased with some good cold of fairly long duration in winter as they know that usually leads to good, even bud push come spring. The reason, I believe, is that that the absorbtion of moisture at those very low temperatures leads to greater "push" pressure because of the greater temperature differential between the cold absorption and spring normal warmth. The pressure is directly a function of the differential at the bud-break stage. More water can be held by wood the lower the temperature (above 32 degrees F). Thus, when heated, greater pressure. This would also explain no bleeding in cold mornings but bleeding occurring later in the day. The "spurting" usually occurs later in the day – a good, warm day.

I mentioned earlier that I was a proponent in the seventies of putting my apples to bed for the winter with wet feet. I believe in the same thing for grapevines. I have personally seen apple trees with very sparse spring performance after a winter with little or no snow. In Europe – with its no irrigation – it is interesting to look at the winter rainfall records and compare them to the wine evaluations, as well as yields, of the subsequent year. Winter moisture available in the soil is a necessity for the vine to absorb to its maximum capability at those low winter temperatures. Water must be available and available at the right time!

Erratic bud push spread over a relatively long period leads to grapes of widely divergent characteristics at any given harvest time. Those sorts of grapes lead to less than desirable wines. Period. There is no correcting in the winery for the green, unripe fruit – even vegetal if you will. Selective harvesting can help but it is very expensive to do in this day and age. With sufficient water available after commencement of dormancy and prior to full winter differential bud break affects will be reduced – not necessarily eliminated but reduced. A warm winter will reduce the *amount* of water for expansion. The stored energy for growth is also a factor, as discussed elsewhere, for the continued growth once bud push is accomplished. But that is not all – the MHP has far more work to do throughout the year. It is *not* just a bud break mechanism.

Here is as good a place as any for a little divergence to set some foundational thought for later discussions particularly in clonal and varietal development but as basic understanding of my thought processes. To many people, if not most, heat and energy are the same thing. To me they are not. Heat is an *effect* of energy and is something we can measure its presence or absence. We creatures even have rudimentary sensing equipment within a limited range – we can "feel" "hot" or "cold" or in between. For a long time there was disagreement between scientists as to whether light was a particle or a wave. Albert Einstein tendered the idea that light was "individual packets of energy" that he called quanta. He had serious reservations, as do I, about wave theory. The generally given example of wave theory was a pond into which a pebble was thrown and the resulting "waves" were like light from a sun. The problem with that is it implies the existence of a "medium" through which the waves are propagated. No known medium exists. Thus, one of his three rules for a theory was violated – namely, the reality requirement. However, there is very useful information couched in wave theory language – wavelength – that should be utilized by the grower (see Smart's "Sunlight Into Wine"). The emission by the sun of "quanta" in a variety of patterns could sort of fit wave concepts and I will use it for what it is. I'll leave theoretical physics to others to argue. We can measure the wavelength and we know that plants use the energy in certain patterns (see "Green Window" by Smart and others).

What has not been explored are the effects, if any, of the quanta patterns in areas other than photosynthesis.

The "packets of energy" idea does have some interesting applications to our attempts to understand the grapevine – granted at only a hypothetical level. We are all familiar with the idea that sun-canes are more fruitful than shade canes (see Winkler, General Viticulture, 1968). We are also familiar with the idea that some light increases anthocyanin density in red grapes and that too much can bleach out color and that some varieties are more prone to this than others. Etc., Etc., Etc.

How might these energy packets affect the vine? Consider we humans when exposed to the sun. At low dosages many of us develop melanin pigment in our skin to protect us – we call it "tanning". Others of us develop no or little melanin and even at low dosages sunburn and blister or even very severely burn. Ten, twenty or thirty years later cancerous melanoma can develop from that sunlight exposure or so statistical studies have asserted. There are other arguments but we'll stay with this. If true, those studies would imply quanta activity at the cellular, molecular or even sub-molecular level on organic tissue. The grapevine is organic tissue. The quanta inject their packets of energy into the cells and their components causing higher energy levels of the atoms and that in turn caused electrons to orbit at higher levels. These excited levels at higher temperatures have all sorts of effects not understood. But certainly a plausible case could be made that in a period of high solar emission enough chaotic frenzy could have occurred such that new varieties were created because of effects on DNA and genes. The same process at lower levels could account for clonal differences – minute differences within a variety. Bud mutation could also have its roots in this mechanism. We all know that Gewurztraminer sprang from a bud mutation of a Traminer vine. The color and the flavor are radically different from the parent. The apple industry is replete with new "varieties" that have sprung from a bud. I'm not speaking of hybrids but of adventitious shoots from a bud.

I also believe that the quanta have an impact upon fruit differentiation. With over-pruning I have experimented with over-loading the buds with available stored energy such that in the spring at the time of differentiation there would be an abundance of energy. The point was to determine if fruit-bud differentiation was a function of that factor alone. It was not. It appears that exposure to sunlight at sufficient *intensity* is a major factor. Thus, the old sun-cane, shade-cane admonition. Notice, I did not say a hot enough day or refer to air temperature. I am speaking of some required level of quanta impact on the developing bud. What that is I don't know.

Earlier I spoke of the Heat Pump having more work to do after bud push. The process continues after bud push. Early on there is not enough new leaf surface to contribute much in the way of transpiration pull. Again, with the cool night spring temperatures and cold soil the vine absorbs moisture pulling it into the roots and up the vine. Remember, during the day the generated pressure was relieved by pushing new growth and its extension. At night re-absorption occurs up to the maximum if soil water is available. And, again, the sun comes up, heats the liquid and pressure results to push growth. At night the absorption "pull" brings elements from the soil and cytokinin hormones from the roots. The pressure system works to boost flow upward to the growing areas. In addition, in conjunction with gravity it also pushes flows downward to feed the roots. The vascular system has a series of small "check valves" allowing flow in only one direction. Pressure operates equally on both regardless of direction. Transpiration "pull" operates in only one direction as does capillary flow and does not explain return flow for distribution of CH to the roots for their wellbeing, growth and energy supplies. Transpiration also does not explain night movement.

The powerful absorption mechanism would also explain the significant pulling of water capability the vine expresses. One common method of monitoring water needs is a ceramic tipped, sealed water-filled probe stuck in the ground. As water is pulled through the ceramic tip an increasing partial vacuum is created and that is read. That is one hell of a force to do that but it does it. That is the absorption component plus the transpiration component during the day. On days when the stomata were essentially closed during wind on the old system the partial vacuum was still occurring which must have been primarily the absorption component.

In most grape areas there is a significant difference between night and day temperatures. There is also a significant difference in air and soil temperatures – particularly at the water pulling depths. Roughly put we see diurnal temperature swings of about 30-35 degrees F. In Monterey it may be 50 at night and 85 in the day. In Fresno it may be 75 at night and 105 in the day. The differential is still about 30-35 degrees.

Long ago plants figured out how to use chloroplasts for access to the energy of the sun. It would be almost incredibly silly or incompetent for plants to ignore such an obvious source of free energy. A 35 degree temperature differential occurring on a regular diurnal basis is too much to ignore particularly combined with water's characteristics of expansion and contraction plus being a good heat absorber.

I have called the process a Heat *Pump* because it acts like an old-time pump. The process alternates: pressure, release, pressure, release. As the day progresses we have all seen leaves lose their turgidity. I think that wood is an insulator to a large degree and that the water deep inside is cooler than the ambient air temperature. Slowly, throughout the day, the absorbed water is rising in temperature and the differential is decreasing. Granted, the cold water from the soil is being absorbed but not fast enough to maintain a high differential. As the differential declines late in the day so does the potential pressure and transference. The primary pull is then the transpiration mechanism which I believe is a relatively weak force. Operating essentially or nearly alone it cannot maintain leaf turgidity in onerous conditions. Finally the leaf closes its stomati to protect itself. At night the onerous conditions are removed and the process starts all over again. Leaf burn and tip burn occurs when the onerous conditions go beyond the vines capability to protect itself.

It is interesting that in severe water shortage it is the older basal leaves that go limp first progressing up the cane. The youngest leaves wilt last. Upon water application it is the youngest leaves that become turgid first progressing down the cane. The oldest leaves recover last, if at all. The same sequence occurred in a small patch of mint I maintain for cooking that I let become stressed for water. The return of turgidity began with the younger leaves within twenty minutes of water and was complete for all leaves within forty-five minutes. The terminal four or five tiny leaves never wilted at all. Under stress, the plant closes its stomati. Thus, it would seem that transpiration pull factor cannot be heavily in play with the water application. Absorption would be a function of mass and therefore one would expect the larger leaves would pull first yet that is not the case. It does make survival sense to focus water on the younger producing leaves. How the plant determines this is a mystery to me. Perhaps as leaves reach post-maturity the vascular tissues close – it certainly is a question to answer. Perhaps it is the weight of the freshly watered soil upon the roots that aids in pressure to push water upward into the depleted vascular tissue. The recovery seems too fast for transpiration pull, capillary flow or the Heat Pump to be the full cause, separate or together. Even grapevines will display signs of recovery within one hour – at least in our rocky soil.

In the past we have seen some "summers without summer" – years in which the normal summer temperatures did not occur. The temperature diurnal differential was much smaller. Thus, not only was the photosynthesis production reduced but the heat pump was not as functional. The following year there would be in the region far fewer clusters and a greater incidence of shot berry. In that sort of summer crop reductions and foliage feeding seemed to enhance our subsequent year performance in comparison to vineyards nearby that did not respond to the coolness in that fashion.

I believe that the heat pump mechanism is the one directly involved in the elongation of the internode cells – not the nutritional aspect. We are told that the vine, during growth at the meristomatic tip, makes a finite number of cells and then makes a new node with potential meristomatic tissue. The number of cells in the internode is fixed and no meristomatic growth can occur there from. That is what we see in practice. We have also been told that elongation of those cells depends upon moisture and "general conditions"! There are those "experts" who have taken the position that shorter internodes at different locations along a cane indicate poor attention to farming – specifically irrigation or nutrition. Once upon a time one such "expert" (*not* from Monterey) was deposed and was prepared to so testify in court. The matter never went to court. However, had it gone to court, I was prepared to testify concerning a phenomenon called The Marble Cone Fire during which the Salinas Valley was smoke covered for about a month and a half. That year all crops there displayed weird things and many people were dizzy and off balance in the strange bronze light. All canes in the valley displayed the shortness characteristic at that time frame elongating better once the smoke cleared.

Let's look at that a little more closely. Cell division and formation requires CH – work is being accomplished and nutrients and minerals are required for that construction. Once they are formed no further construction materials are required. The existing cells simply "elongate". It is obvious that the amount of elongation is a variable depending upon something else. I am of the opinion that that "something else" is the Heat Pump. In the Marble Cone Fire example sufficient plant usable light was getting through the smoke for cell division to take place but not sufficient for normal, higher temperatures for that time of year. Thus, the temperature differential was smaller and, combined with reduced transpiration loss, the total system pressure was reduced. That reduction of pressure meant that there was less pressure available to push water into the progressively more resistant internode cells. One reaction to conditions that result in shorter internode length because of coolness or smoke or volcano debris should be to lengthen the irrigation interval. Remember, grapes don't like wet feet and root-cell deterioration can commence quickly – assertedly within 48 hours of saturation. The vine is using less water in these conditions and one shouldn't try to "force" growth with irrigation. Also, one is further reducing the differential! That calendar determined schedule should be abandoned as soon as the situation is noted. The approach of trying to force (which I have seen attempted often) actually exacerbates the problem and can leave a damaged root system to face a later hot spell.

Another consideration for the need of a "pump" is the circulation system need to push nutrient laden liquid through small channels to the individual cells then picked up by a return flow. In we humans the arteries carry blood out, it goes through tiny capillaries and is picked up by veins returning it. The pump (heart) pressure is on the artery side. In the vine the Heat Pump pressure is equal in all directions however, during the growing season, the transpiration loss of the leaves creates a pressure leakage on that side. Thus, there is a differential for the flow across. One might rightfully then ask about the transference across the other way. Remember the "check valves" in the vascular channels. At night the absorption factor can only flow liquid in the up channels. With the absorption the wood tissues expand creating more pressure on that side of the system. The pressure so built is felt across the vine and forces flow downward - the only direction it can go because of the check valves. That creates a mini – differential that the higher pressure tries to equalize by flow across the capillaries. With sunlight the cycle begins again.

Understanding this, even if hypothetical, will help the grower not to panic and run to irrigation and fertilizers. These situations all pass with a little patience. Cold spells go away, smoke clears, volcanoes go back to sleep, etc.

The Einsteinian "energy packets" concept of light and the associated absorption thereof by the vine seems to provide a reasonable explanation of many activities of the vine we observe in the field. Capillary pull and transpiration pull do not give us satisfactory answers because they are unidirectional. Absorption absent the cycle energy/heat aspect fails to answer the needs of a full circle circulation and distribution system. Whatever the actual truth of the matter, this working hypothesis has functioned well in practice this last some thirty years for me.

Another practice I introduced was the shoot tipping practice to try to achieve cane length uniformity. When we tip a shoot but not others that are shorter we see a bit of bleeding at the cut location. That is the normal healing process. But for a period of time that shoot has lost its full pressure system and has no tip to push nor the ability to do so. It does not even begin to try to push side buds for some time – until it has sealed up the cut. Meanwhile, the shorter un-tipped shoots have their pressure system and their extension actually accelerates showing slight increases in internode length, climate allowing. We can see the related affect from reduced pressure in spring and fall internode length – the effect reduced pressure system (because of reduced differential and photosynthesis). The vine may actually use the Heat Pump system as one of its indicators that winter is approaching.

Another commentary along this line is the age-old technique of handling "frost-pocket" vines before frost protection water systems were devised. Still used today, the technique involves pruning at normal time to a 12 inch shoot. When bud push begins one comes along and prunes to two bud spurs. The industry always referred to the "shocking" of the vine causing about a two week or so delay in the pushing of the remaining two buds. I do not think that we "shock" the vine in the trauma sense. I think the pressure system has been opened and the vine must now heal and seal before it can re-pressurize and get on with it.

There is an element of location sensing by the vine for push preference. An aspect of this "location sensing" by the plant can be utilized. For example, both Riesling and Cabernet Sauvignon have a strong tendency, when a cane is used, to not push buds in the middle of the cane. By taking the cane up over a wire (located about a foot above the cordon wire) and bowing back down to tie off at the cordon wire will result in those mid-cane buds pushing better. The apical-bud business is at work for us.

It also seems that as the early buds begin to grow they are preferentially tapping off a little of "x" energy such that there is little left to push other buds unless the total bud count is low enough to cause even growth. It also appears that the growing tips generate a hormone that suppresses the growth of those below them in their own little mini-system. This may be the method the vine uses to assure its "climbing to the sun" needs. Further discussion of "x" energy will follow in the Pruning section where it is particularly germane.

If one peruses the various old texts and writings one will regularly come upon the phrase "period of grand growth". This refers to a period of very rapid cane growth that occurs after bud push in the spring. Eventually, the writings say, this rate will decline and much slower cane extension will occur, if any, at some point in time. This period of "grand growth" occurs from two sources. The first source is from *stored* carbohydrates that the vine manufactured the *preceding* season. As the new young leaves form they are using more food than they are manufacturing. In fact, it is quite some time before they become positive contributors – that is, making more food than they are consuming in their growth.

To my mind, this is a critical period for the winegrower to consider. One should be trying to build the productive canopy rapidly then hold – a technique I call "mold and hold". The initial energy for this is purely stored carbohydrates –

which I call "CH" cleverly. It takes a lot of effort to come up with these notations! When I came into the profession in 1972 the practice in the industry – everywhere that I saw or was aware of – was to cut off any irrigation at some early point, ride it out through harvest with desiccating leaves on the vines and call it a year. Complete defoliation would occur soon forcing the vines into dormancy, winter rains would eventually come and there was no point in paying for electricity or diesel fuel – whichever operated your irrigation system. For the grapegrower the year was over.

There are several problems with this approach even though it was the industry standard and still is in many minds. If you doubt this go to many of the old books with pictures at harvest time. Look at the color of the leaves while the harvest is underway. The ones from Europe are the best source but extant ones of California are around. It wasn't just the grape growers. In the tree fruit industry there were debates about "going to bed" with wet feet or dry. I happened to be on the "wet feet" side of the discussion.

It occurred to me that as near as I could tell the only source of "CH" stored in the vine would come from green leaves photosynthizing! Yellowing or desiccating leaves couldn't be a very effective source of "CH" if any. 1977 was Ventana's first commercial harvest. Following the last gondola coming out of a field I turned on the sprinklers judiciously in the block. I wanted to hold those remaining leaves green and functional as long as I could. One doesn't want to give so much water that a quirky hot spell could cause new growth – just enough to keep them green. Depending upon the variety, harvest time and year that period of time from harvest until winter forced dormancy could be 30 to 60 days of "CH" production! Notice that perhaps even another irrigation may be required! Eventually day-length and temperature will force dormancy and defoliation. Of course, this grower had cane lignification accomplished before harvest. This is just a matter of maintaining but a very important one – ONE IS NOW FARMING NEXT YEAR'S CROP! Yes, this crop may be over but the winegrower is now into the next cycle.

The vigor and performance of the next spring's growth is, within weather variables, completely a function of the accumulated "CH" within the vine. Notice that, in optimum conditions, the crop load has seriously depleted the vine of its "CH" load – essentially little should be in storage at the time of harvest. This will make sense a little later. At this depleted state, absent a "re-charge" period, the subsequent year's growth will be reduced giving the vine the appearance of low vigor or "overcropped". The grower will respond, over time, to this chronic condition by reducing buds at pruning. These *repeated cycles* will lead to an overall reduction in potential crop that the system *could have* produced! In other words, that industry-wide practice led to a loss of *potential* proper crop on an industry wide basis *within the existing farming structure*. If one simply thinks in terms of *total* produced "CH" in one season the conclusion is obvious. Sun to leaves to "CH" to total fruit is the sequence. Scratch out the leaves for part of the season and the "total fruit" part of the equation is a lesser number. The more "stored CH" one can generate the better the next year's *potential* crop.

The old maxim of "dormant pruning is vigor inducing; summer pruning is dwarfing in nature" is accurate. Green pruning removes leaves – CH producers. Allowing leaves to senesce reduces produced CH. Period!

When I commenced post-harvest irrigation in 1977 I was questioned by other growers and extension agents about "what the hell" I was doing. "Castigated" might be a better term! My "IQ" and sanity were issues mentioned here and there – a pattern that was to repeat itself often. Eventually the practice became more accepted and in the 1980's there were papers well-written by our Viticultural extension agent – Rudy Neja – as the subject and the practice recommended. The post-harvest root growth surge also finds moisture hospitable.

There is another aspect to this "late season" irrigation. Once the winter pressures of day-length and cold have taken over I would lay down another irrigation. I wanted to "pre-charge" the soil moisture reservoir. This is part of the "going to bed wet" thing. In most all irrigation waters there are "salts" of some sort or another. When one irrigates one puts down water and the vine extracts the moisture in the form it wishes leaving behind the "salts". This process goes on over and over leaving a potential salt build-up in the soil root zone. Vines do not like salts and they are detrimental to the soil in the long-run if allowed to accumulate. By "pre-charging" the soil reservoir the winter rains will then be able to "flush" the salts cleansing the soil somewhat, the amount varying according to the quantity of rainfall and type of soil.

The second part of this "grand growth' period comes from the new leaves that have reached a maturation such that they are now *positive* contributors to "CH" production and, thus, providing a new source of food for growth. It is readily apparent that the sooner and fuller those leaves are developed the sooner they contribute. That is, we are maximizing our production of "CH" on a seasonal basis. Up to some point here stored "CH" is the declining source and the new "CH" production takes up the burden. Notice that in a depleted, or nearly depleted, stored "CH" condition development of sufficient new "CH" can be severely delayed. The vine will appear "stunted". Again, we can look at a representation of this concept as CH/B where B is the number of buds left and CH is the total *stored* carbohydrates from the years before (notice the plural).

When does the period of "grand growth" end? And why? For the sake of display let's first assume a balanced vine – proper shoots, stored CH, etc. At some point the vine decides that it is time to bloom. It does this and, hopefully, "sets" its crop. In some varieties this is *not* a given – Grenache is a notorious case in point. Once the berries start to grow the seeds begin to generate a hormone that tells the vine "Send all the goodies to me because I'm reproducing the species" or something like that – maybe not quite so pushy but still it is an order – not a suggestion. Normally in nature species out ranks individual and that is the case here. Just as a human fetus takes hormonal control over the mother's body and will take calcium from her teeth if not enough is available so the grapeseed will exert control over the vine's allocation of "CH". As the seeds generate more and more control taking more and more of the CH the amount of CH remaining for shoot growth diminishes. This is the "sugar accumulation system" – *sugar* only.

Consider the situation where too many seeds pollinate for the vine's condition – either not enough stored CH or new CH – and we can see that the cane growth will come to a screeching halt at whatever length it had at the time of seed-formation. Grenache is, again, a good example of this in some years. The usual – and correct – grower response immediately after set is to make a thinning pass – green dropping the excessive clusters on the ground. This is NOT the same thing as adjusting for quality later as that must not be done until veraison or after. The power of those seeds to debilitate the vine is awesome to those who have seen it. I have seen many, many crops around the countryside brought to harvest with 12 to 18 inch canes and dehydrating the fruit down to sugar. Those wines must have been delightful! SEE PLATE 9, 11 12 and 13.

Now consider the opposite – not enough seeds on a vine to control the vine's growth. This one is a little trickier and involves an area that I'm sure will be a little more contentious. At first blush it seems very straight forward – no control by the seeds and the vine can/will run rampant. One might say "so what". That is exactly what we saw in American vines in the past – canes running across the ground areas of twelve foot alleys!

I must segue into another discussion and then will return to this part. The vine prepares itself and the fruit for coming winter in several ways. For one the vine is somewhat day-length sensitive and will begin slowing its growing processes as the days become shorter and shift to "ripening" processes – both wood and fruit. Another is soil moisture availability that can stop the tips of the canes from growing. Another is the crop load that can slow down or stop the tips of the canes from growing. The common thread through all these is the TIPS OF THE CANES.

In a shoot the only active meristomatic tissue (tissue where cell-division is taking place) is at a growing tip. The other tissue is fixed. That is, the growing tip manufactures just so many cells and then manufactures a "node" or point that contains potential meristomatic tissue. After those internode cells are formed they never cell-divide again and go through a period of elongation which is a function of available moisture and Heat Pump pressure.

Nutrients have a different role in all this. Those "tips", in the act of cell-dividing, generate another hormone- a "suppressing" hormone. This "suppressing" hormone CANCELS out another hormone generated in the seeds that tells the plant to form the "ripe" compounds. It is separate from and operates independently of the "sugar" accumulation hormone.

Why would there be such a thing? Well, the tips are out there in the world. They can feel that the days are still long, they feel the sun shining and they are getting plenty of moisture. They know life is great and its summertime and "livin' is easy". They keep sending down the message "not yet, not yet. Winter is a long way off". But when things change and that tip is no longer rapidly cell-dividing less and less of the canceling hormone is being sent down to the berry. Also, reduced circulation in the vine will result in reduced canceling hormone being delivered which is a factor when growers start dehydrating down to sugar Brix or restricting water. The various hormonal systems operate independently but also as an integration to varying degrees. They are designed to function on their own but human manipulation of factors such as nutrients, light and heat will harness them for our goals! For display purposes, there is an inter-relationship of parallel operation in nature – each affecting the other according to temperate zone conditions. Perhaps it is better to say that they are "complimentary" systems. The information encoded in these hormones could go through a variety of transformations based upon a changing frame of reference but the basic laws governing the relationship of the components to each other remain the same.

Why would one suppose and hypothesize the existence of these two separate and independent systems? I will display two examples. The Riesling grapevines we have here are, of course, from Germany. You can't lay the "soil" thing on me because I have personally seen Riesling growing in every type of "soil" imaginable in Germany. In California the old standard for harvest was 21 degrees brix but many wineries pushed for somewhat higher towards 22 degrees. In those days (you will find this in the old texts) a standard rule-of-thumb conversion rate of sugar to alcohol by volume was .56. In my opinion, in order to have a chance to make an outstanding Riesling wine the grape must taste of apricot! American vineyards could not get this (if at all) until beginning around 22 degrees. Most Riesling wines then were wet, white and sweet and of no real aroma unless doctored with Muscat. Thus, dry wines from grapes in this range (21=11.76 ALC; 22=12.32) were barely into proper "RIPE" compound formation. Comparing those general California Rieslings to German ones was (and is) a shock! Wines of 9, 10 and 11 degrees alcohol that are dry (TROCKEN) with lovely and delicate floral smells are common. Of course, these have little "body" but they are not so intended. A 10 degrees alcohol dry wine would have to be from grapes harvested at 18 degrees Brix! How in the world could that be – "ripe" flavours at that degree?

During the period 1978 - 81 I was deeply into the search for the Holy Grail – Pinot Noir. In Burgundy winegrowers are authorized to chaptalize – the addition of sugar during fermentation up to 2 degrees alcohol maximum. When I would ask Burgundian winegrowers what they did in those years when they had sufficient sugar in their grapes they would answer "Oh, I have to add sugar of at least one-half degree alcohol to make my wine". Typically Burgundian wine will be in the 12.5 to 13.3 alcohol range. If we back off the one-half degree sugar alcohol from that then we see 12.0 – 12.8 natural grape sugar. These translate to 21.4 to 22.9 degrees Brix at harvest. If we pull off the full two degrees "sugar in a bag" then the corresponding numbers are 18.75 to 20.20 degrees Brix at harvest. Can you believe that? Wow! Yet the wines taste and smell not vegetal or un-ripe! How can that be? How can there be fully ripe flavours at these degrees sugar?

The observation that "ripe" flavours could be developed at significantly lower sugars than we experience here meant that there had to be a different controlling factor than the control for sugar accumulation. Pretty straight forward. But where was it?

If one sits down and studies the various pictures in the books some things become rather apparent. The Burgundian and German vineyards don't look like ours! For instance, the Burgundian vines are roughly waist-high but that's not the important point. There are no canes sticking up or out! No tips. They've been cut off! Same in Bordeaux. Same in many other places. In others they've just stopped growing. Revelation. If one cuts off the meristomatic tip then for awhile no "suppressing" hormone is being generated and the seed-borne "ripening" hormone takes over. As side shoots form – if they do – they begin generating the suppressing hormone at their tips and must be cut again. We began snipping tips by hand in Riesling in 1978. Efficiently shearing tips mechanically at the proper time and length did not begin until we had vines in the new system held upright by wires. Then it is pretty easy.

Other methods can also be used to slow or stop the tip growth. One is irrigation. The Ventana is basically rocks and sand so response to water changes is quite rapid, comparatively. One old time trick for evaluating a vineyard's moisture condition is to look over the top of the vines. If the sensation is "tendril" – that is, the tendrils are longer than the tips – then the vines are in luxury. If the sensation is "tips" then one will need water before too long. This characteristic demonstrates that the vine undergoes physiological changes as conditions vary. Water shortage can pull the shoot extension to a stop. The cell-division can slow to nearly nothing or stop thus failing to make sufficient suppressing hormone.

Another method – the one I prefer – is to use the crop size to put pressure on the vine. The other two are available, if necessary, to compensate for misjudgments of a not-enough crop situation. The use of water is really no different than the throttle in a jet fighter – ease it forward for a little more, ease it back for a little less. There is no requirement that irrigation run exactly 12 hours or 24 or 36 hours. Vary it according to conditions. The ideal crop size, to me, will be the one that pulls the tips to a stop naturally about 30 – 35 days before harvest with, of course, properly developed canes of 14 – 16 full sized functioning leaves per shoot.

I view "cane-cutting" or "shearing" for tip control as basically "remedial" in nature to correct for other errors causing excessive vigor such as too much fertilizer, wrong rootstock for spacing, wrong spacing for variety or improper pruning philosophy. Sure – it can make the vineyard look very pretty with its hedgerow trim appearance but one should ask why all the vigor. The clue is there – the vine would like a little more crop.

There is a wrong way to approach "shearing" or "hedging" except as a facilitator for harvest machine access. One example is a presentation at the Symposium on Grapevine Canopy and Vigor Management by Dr. Mark Kliewer of U.C.Davis in August, 1986. He showed a slide of the cutting process wherein vines that had grown down and partly across the alley were being cut such that 50% of the sun-exposed canes and leaves were dropped. Of the remaining leaves the heart of the vines in this old sprawl trellis had already senesced. From this he had concluded that "hedging" delayed harvest by two weeks! After his presentation I spoke with him pointing out that he had cut off most of the mature CH producing leaves leaving senesced heart of vine leaves and the oldest, over-age leaves. It was my opinion that he didn't have 30% of the required leaves left. In all probability the harvest Brix was achieved by dehydration and/or sugar transfer from wood. This "test" was in 1983, 1984 & 1985. Each year they cut the same-vertical at the shoulder. This wrongful conclusion persisted within the industry for years. The shearing concept should not be used in this fashion at all. At the most in this situation "skirting" (that is, the cutter bar used horizontal not vertical) could be done – but much earlier. The problem in the subject vineyard was that the vineyard itself was out of control and poorly farmed to begin with. Whose it was I do not know – it could have been one of Davis' own. Tip shearing should be just that – cutting off tips.

Given this background diversion let's return to the situation where too little fruit is on the vine to pull down the growing tips. In the picture just noted above and in most other vineyards the conditions covered were usual in the warmer growing areas such as Napa, Sonoma, Livermore and the Central Valley. Monterey's vines were too young to display the pattern but later many did. However, many others were more characterized by a pattern of early delayed growth due to winds and cold later achieving their growth, such as it was, very late in the season still growing as they were harvested.

Shifting back to where we segued into the above background discussion we can continue the trickier aspects of under-cropping. A slight under-cropping poses no particular problem that some more diligent farming can't solve such as shearing, reducing water, reducing nutrients, etc.. However, even slight under-cropping combined with inattentive or unknowledgeable farming can have terrible effect upon the wine. If the crop does not have sufficient mass to pull the tips to a stop and the vines are left to grow at will they are continuing to produce the canceling hormone for the "ripening" process while the "sugar" accumulation process continues. Therein lies the problem. Grape contracts between growers and mid-sized to large wineries are based upon degrees Brix. I think I wrote the first contract in the industry with pH and acid mentioned as parameters in either 1979 or 1980. Over the years we have been primarily a grower with a small winery for study purposes and as home for any grapes excess over sales. I have had many, many small winery owners/ winemakers and home winemakers taste the grapes when told the numbers are there. Get this – I have NEVER had a representative of a larger winery EVER TASTE THE GRAPES AT HARVEST TIME! Test – yes, taste – no. Change that little "e" to "a" and you have to make a judgment. Such things are not appreciated in the corporate world. They are detrimental to one's employment. Numbers are safety.

Degrees Brix is a nice easy measurement, very simple and somewhat accurate – good enough for wine work. Sadly, there is no easy simple measurement yet for "ripe" – sugar, yes; Ripe, no. None, that is, except the Mark 1, Model A human mouth and nose and experience. I learned very early on that one indicator that "ripening" was possibly occurring was browning of seeds but it was not an absolute. However, the opposite has some merit – no brown on the seeds means we are not in the "ripening" zone.

To explore this idea of two major hormones involved in "ripening" versus one hormone controlling sugar accumulation, in1980 I deliberately lightened the crop on a few Chardonnay vines near a waterline valve where I could access water. The vines were maintained in a growth mode until the grapes measured 29 – 30 degrees Brix by hydrometer – the common tool at the time. The grapes were turgid – not raisined down to sugar. They were grown under canopy and thus were greenish not yellowish. By taste these grapes were unripe with greenish flavours like unripe apples (probably from high malic acid)! When I spoke of it I thought everyone would go "Wow – that's interesting". Nobody gave a damn! I had demonstrated what I thought was a very important point for winegrowers. The reason that this is such an important point is that really the objective of the whole thing is to achieve the intersection of the two systems. That is, achieve the desired ripe flavour compounds at the desired sugar.

These hormonal systems may seem a little convoluted but having a working hypothesis that explained observations was necessary for developing a growing system that addressed all the issues that were known to us then – in the seventies. Of course, there are others doing various things in the vine and berry and I leave those to young minds to sort out. These are the ones that were important to me to acceptably solve Monterey's cold climate problems and, in passing, some of what I perceived as California's problems.

I believe that balancing these ideas has led to wines made from Ventana Chardonnay grapes receiving gold medals for 29 consecutive years and those from Riesling grapes now 30 consecutive years! Recently, at a "Chardonnay Shootout" held in Sonoma for all of California the Ventana 2005 Chardonnay was judged the Best of California. How else to explain it? That "consecutive years" stuff means that we were able to use variable FPs to counteract the C variable.

Ideas or Hypotheses are tested by the consequences
which they produce when they are acted upon.
—John Dewey

Section 3F
YIELDS

Quantity per unit area (tons per acre) is the common terminology *and mindset* within the industry both here and elsewhere in the world. In terms of accounting, when the current
game is over, I guess it is the easy way to "count beans".
Unfortunately it leads winegrowers into thinking in those
terms and from that viewpoint. For a grower of fine wines
– or even medium wines – the only relevant mindset should
be in terms of pounds per vine within a given system for a
specific use.

"Balance" in a vine between root, canopy and fruit is the
factor one should be seeking. From a "Platonic ideal" approach one should be seeing a crop load that results in a
canopy growth of about 14 – 16 mature, expanded leaves,
grown on time, and that can essentially stop tip growth a
month or so before normal harvest date. It would be lovely if
we could achieve that in every vine in a block but I've never
seen that nor accomplished that. It is that goal one strives
for but it is not given us to reach.

Changing a paradigm is always a challenge. Someone
once said, paraphrased, that the most dangerous area on
earth is the arena of ideas. There was a time when I would
have argued that an aircraft carrier deck was but having
experienced these last thirty-some years as a proponent of

change I would have to agree. The resistance to a paradigm change was awesome particularly if it didn't come from the conventional academic sources. In practice, the academics *follow* the practitioners then claim leadership of the herd!

During the seventies and eighties Napa County averaged around four tons to the acre on Chardonnay and Monterey around three and a half. Many premium winemakers had it in their head that *two and a half* tons was the "ideal" or "proper" load for their needs. Even today there are many winemakers who push their growers to reduce crop loads to ridiculous levels - even at contracted levels. Those "contracts" are written, usually in advance of the growing year, often years in advance. These contractual yield limits are "boilerplated" into generic contracts with no consideration of the vineyard systems – one size fits all! The "winemakers" and "Reps" of the medium to large wineries are usually young graduates in "Fermentation Science" *not* "Viticulture". This leads to a lot of problems in communication and understanding particularly when the person with the checkbook is operating under the belief that myths are facts!

There is a deeply held myth that "the lower the tonnage the better the quality". Various writers, critics and publications firmly believe this over-simplification and continue to spread this myth as if it were on a third tablet brought down from the mount by Moses – a dictum from God that cannot be changed. What complete and utter nonsense and a terrible dis-service to consumers and industry alike. I have already shown – in the hormone section – how an undercrop situation can/will lead to sorry wine with unripe flavours and taste. Tons per acre at some mystical and arbitrary number is no indicator of physiological development of the grape.

Consider now the historical tonnage numbers given above for Napa and Monterey. At that time the extant vineyard design prevalent in both areas was 7' x 12' which gives 518 vines per acre. Napa's 4 tons equals 8,000 pounds and, at 518 vines, equals *15.4 pounds per vine*. Monterey's 3 ½ tons = 7,000 pounds which equals *13.51 pounds per vine*. As

I mentioned earlier, in 1978 and again, larger, in 1982, we installed vines in our new system at a spacing of 6 feet between rows and 3 ½ feet between vines in the row. This is a vine-count of 2074 vines per acre. If we assume a yield of 8 tons per acre = 16,000 pounds divided by 2074 we see a yield of *7.71 pounds per vine*. Compare that 7.71 to the Napa 15.4 and Monterey 13.51. If the mantra of lower tonnage is better quality is true then a rational winemaker should desire the 7.71 grapes. Not so – the 8 ton per acre number was so far beyond their myth belief and what they had been taught that they simply could not comprehend it. It took more than *twenty years* to be digestible! A few years back I had a young field rep for K-J stand in front of me bragging that many of their fields had 8 tons and over of quality fruit! I simply listened and smiled. The revolution had occurred!

It may appear that I am talking out of both sides of my mouth when blasting the lower tonnage idea when the above vine numbers indicate that. No so. First, it could be that the prevailing system was over-cropping the vine – maybe or maybe not. We did see a radically improved grape and wine quality which the theories predicted and would support the over-crop idea. However, the quality shift was, I believe, a function of the *entire new system* at work – light penetration, exposed leaf area, etc.

Perhaps twenty-five years ago in Sarasota, Florida at a wine weekend sort of thing I and Phil Woodward of Chalone and a couple of other folks were on a panel. I was discussing aspects of this new system and, of course, blasting the lower tonnage business. Afterwards, Bob Mondavi cornered me (others gathered around to listen to us) and took mild exception to my contentions, asserting that they had done wine evaluations relative to reducing crop loads. The lesser loads gave a better wine. I asked him if that was all they had done – just lightened the load and he said it was. Mondavi, in those days, was doing important research and experimentation in winemaking *not* viticulture. I suggested to Bob that his observation on yield/wine would indicate an initial over-crop situation within his *system* within the vineyard. Mondavi was, of course, using the Davis "hay stack" system. We had a nice long discussion about changing the *system* and its effect upon the wine. Bob later told me that that conversation led him to focus more on the vineyard.

As time passed and the larger planting came into full production in the later eighties – the rumors were spreading that I was yielding 8 to 10 tons per acre on Chardonnay. It simply wasn't true at that time but that didn't matter – I lost quite a few grape deals because of that belief. The small wineries – well, most of them – continued but the bigger guys bailed out based upon the rumors. The humorous part was that I simply made wine and sold it to the big guys for far more money! Winemaking is not terribly complicated if starting with beautiful fruit. Bulk wine purchases are made based upon the perceived *quality* of the submitted wine samples not the mythology of yield! I have *never* been asked by a bulk wine buyer about yield per acre on a given wine! Big winery or small – never! On the bulk market the wine speaks for itself.

Historically, one concept detrimental to this industry is the huge cleavage between viticulture and enology. At the university level the cleavage exists between the two degrees "Fermentation Science" and "Viticulture" and there is little cross training. The enologists view themselves as the "scientist elites" and the viticulturists as "farmers" or "peasants". Historically, it was severe at the time I came in to the industry. The two groups barely spoke to one another and, if they did, it was usually padded with tension. Growers were not members of the elite club – their job was to grow the grapes to the assigned degrees Brix and accept whatever price the winery decided to pay – often decided after the fact. As support for this observation as an outsider I point out that in my early years here the prestige organization was The American Society of Enology. Only recently has "and Viticulture" been added! The growers felt they were cheated by wineries at every turn and perhaps they were. I've heard some pretty horrific stories. I've experienced a couple myself. Laws are usually passed because of onerous conduct in the marketplace. Consider that now, by law, State Inspectors are required to take sugar readings at receiving stations if the grapes are being purchased. If the sugar readings before were honest why the demand for State involvement?

I bonded our winery in 1978. In 1979 or 1980, a meeting was held with winery owners and/or winemakers to discuss putting on static displays of certain problems in wines at a Monterey County Grape Growers assembly. I suggested that we demonstrate the problems with pH because I had been deeply studying this and thought it very important. A well-known winemaker leaned forward and said "Oh, Doug, we can't do that. These growers will never understand pH". I don't know what an "umbrage" is but I took umbrage – whatever it is – with that remark. I said "Well, I'm a grower and I think I can handle the concept. So can they!". Another person at the table – a winery owner and grower – said "Doug, you are now a winery and you better start thinking like one."

The detrimental affect of this "cleavage" is that the two camps tend to live in their own separate worlds. I think that fine winegrowing is just that – growing wine. I think that it is a continuum not two separate activities. To the extent that parts of the process are adversarial the final result will be impaired. So many decisions made in the field have repercussions in the cellar and in the glass yet usually those decisions are made by people with no idea of winemaking. There is literally no feedback to a grower on how to change practices to improve the wine. In 1978 I bonded the winery for a variety of reasons not the least of which was feedback to make me a better grower. This feedback is crucial particularly so if you subscribe to the idea that you need a great grape to make great wine. The best of all possible worlds is where winegrowing and winemaking is under one control but not large corporate. Look to the old world's pattern and its results. The Viticultural procedures reflect the feedback

from the cellar. To this day, with some enlightened excep-
tions, I see wineries sending as their "field reps" to vineyards
people who are fresh out of school with absolutely no idea of
viticulture – none. Often they also have no idea what these
grapes are going to be used for! They seem to have one mission only and that is to look for any diseases. It is incredible
to me that at the one interface between these two halves
people of the lowest knowledge and experience level are
chosen by wineries to represent themselves.

This cleavage – to this day – can be demonstrated by two
completely divergent paths taken by the two halves. Some
of us, myself included, have led growers into techniques that
will lead to greater flavours at lower sugars. That is, getting ripe flavours more and more pronounced and reducing
off flavours (more uniformity) at table wine sugar numbers.
Meanwhile, most winemakers have been pushing for higher
and higher sugars from growers. It seems like they are trying
to out-Parker Parker or the Wine Spectator taking table wine
alcohols into the stratosphere – where, in my opinion, they
don't belong.

Or it may be in some houses that a little greed is going
on – pushing growers up high on sugar by dehydration (thus
lowering the weight across the scales by as much as 30%)
then adding water back into the grapes at the winery now
that the law has changed to allow that (pushed for by wineries and fought by growers). Wineries make more political cash donations than growers. You see – water is much
cheaper than grape juice. Many growers believe this is true
and are angry about it. The longer grapes hang the greater
the risk of fall rains and the grower doesn't get paid for more
risk. Also, there is little or no storage time for "CH" for the
next year's crop. Growers feel, rightfully, that they are getting
hit by shrinkage and bird loss in one season and that their
vines are being debilitated for future seasons. Winemakers
simply don't understand this – or don't care. It is that lack
of understanding or caring that has growers so upset. The
internal wars go on.

Later I will discuss the effects upon yields as a function of machines, pruning and design interrelationships. These are all secondary to the understanding that usable fruit is a function of generated carbohydrates per unit area. Maximize the CH and you find the yield limit.

All thought is a feat of association: having what's in front of you brings up something in your mind that you almost didn't know you knew. Putting this and that together. That click.

—Robert Frost

Section 3G
IRRIGATION

In general, the western Europeans have taken a necessity and made it into an icon. Their history of winegrowing stretches back at least three or four thousand years and 8000 years for the Middle East as far as we know. They have had plenty of time to match soils to prevailing rainfall patterns and to winegrowing varieties and procedures.

Here in the new world we are very much the newcomers and we are facing significantly different conditions than those of the old world. One huge difference is that our continent has mountain ranges running essentially North and South while Europe's mountain ranges are east and west to a very large degree. It may not seem like a very big difference but it is – a very big thing in the movement of peoples and plants historically. In the old world it allowed the relatively easy movement along common *latitude* lines thus not a great change in conditions. Further, it seriously created different climate conditions. In the northern hemisphere the cyclonic air flow is counter-clockwise. The storms off the north Atlantic carrying water are channeled deep into Europe by those east-west mountain ranges which themselves create eddies and vectors that act like billiard balls bouncing around. Here and there, there are smaller north-south ranges like the Vosge and others which further create microclimates, rain shadows and the like. The moderate elevation rises cause rainfall from even regular mild fronts from the north Atlantic airflows. These conditions allowed Europeans to find locations that were generous to various crops according to their water needs within the prevailing general rain pattern situation. As we look throughout the world we can see that irrigation was not an unknown concept – the ancients practiced it where needed. Babylonia, Egypt, China, Maya, etc., etc., all practiced it. But not in Western Europe on any scale. You might ask about the canal systems throughout various areas of Western Europe but I point out that those canals were developed *not* for irrigation purposes but for commerce. They are not really used for irrigation purposes even today.

In the new world (North and South America – I'm excluding Australia) those massive north-south mountain ranges of the western side of the continents create substantially different conditions. The cyclonic flow of the north Pacific impacts those barriers causing a rapid climb in altitude of the air and the resulting dumping of rain on the western side. Once over the mountains the new somewhat depleted air mass descends and, voila, a rain shadow upon the land. You can readily see this affect whenever one drives over a mountain pass – lush forest and big trees on the west side – sparse forestation and smaller trees on the east. Obvious.

To further compound the difference the north Pacific Ocean is just that – generally pacific – in the summer months – at least as far as we are concerned. We can go very long periods with no effective rainfall at all – consider 1976 and 1977, for example. Grapevines do not like sodden conditions nor do they like rainfall as the fruit is ripening – well, maybe they do but we don't. They surely don't like conditions where their feet are in saturated conditions when they are in the growing stage. During dormancy they can tolerate saturated soil with no cell deterioration but during the "green" season cell deterioration can commence in 48 hours in a fully saturated situation!

Those North American mountain ranges also created other "problems" that our ancestors had to sort out. The east-west movement of peoples was *very* difficult. In my past I regularly transited the United States from California to Florida in a fighter – out and back in the same day. On clear days looking down I never failed to be in awe of those old folks crossing the country in wagons! Absolutely incredible people! During the Gold Rush period when modern vinifera wine grapes first arrived in California (I'm ignoring the mission grapes brought by the padres) most of the "49ers" came by ship "around the horn" – a very long and arduous trip in rickety wooden sailing ships – another feat that awes me. I've been on replicas and I can't imagine being in one of those for 60 to 90 days in the ocean. Aboard an aircraft carrier I've seen eighty-foot waves in the Pacific and our escort destroyer going submerged then re-emerging then submerged, etc. I just can't imagine how incredibly tough those old-timers were to be out there in tiny wooden boats – or how nuts! Peter Mirassou likes to tell the story about his ancestor Pierre Pellier who brought his cuttings from France aboard a sailing ship that ran low on fresh water. The captain would not allow him water to moisten his cutting so he bought a bunch of potatoes, of which they had plenty apparently, from the captain and stuck his cuttings into the potatoes. The cuttings survived. Again, the power of absorption is demonstrated!

Some of the "49ers" came not to mine for gold but to "mine the miners". Maybe they were originally thinking gold but soon learned that that was very hard work or not to their liking. Ones like Pierre Pellier obviously had growing grapes in mind in advance but others had a change of mind once here. The center of the activity was San Francisco with Napa and Sonoma within reasonable proximity.

It didn't take some old folks long to recognize the merits of those areas for raising some grapes, making some wine and selling it in San Francisco – particularly those from the southern areas of Europe where rainfall was more scattered out. The idea of giving each vine plenty of space such that the vine could withstand long periods without rain was not new to them. Even so, a lot of experimentation of matching vine to soil to rainfall occurred. That wide spacing of vines became the norm. It was the method in many areas of southern Europe such as the plains of La Mancha, southern France and areas of Italy and Greece. It still is today. The mind-set became entrenched in the American perception because of the necessities of the area for the next hundred plus years. The big changes occurred in the Central Valley with its hot climate, relatively flat land and water sources flowing from the Sierras that allowed the development of irrigation canals for distribution of water and "flood irrigation". The center of California winegrowing shifted to the Central Valley with its jug wines at much less expense. The Central Valley vineyards could yield far greater *quantities* of grape per acre. Quality of wine – as we *now* understand it – was not a factor. In fact, if you peruse all the old writings from extension services, schools and professors of the subject you will find yield, yield, yield and "apparent disease –freeness" as the recurring commentary! Those were *the* criteria for selection of plant material! By the way – that is a recipe for clonal selection. I have found *not one* reference or comment on wine quality therein! These selections are the plant material the industry had in hand as we entered the 1960's. With the advent of Prohibition many of the Napa and Sonoma wine operations failed. A few survived because of their connections to the Catholic Church making sacramental wines. Post prohibition recovery was in predominantly the Central Valley. The funding of U.C. Davis and Fresno State came from the Central Valley operations and was focused upon *their* problems. The coastal regions were really unimportant to the big scheme of things. There was a dinky "station" at Oakville but judging from papers and reports generated on the work there nobody could remember its name. It was really low on the priority list – almost on afterthought.

One must keep in mind the "geist" of the time – the cultural mind of America. Wine was not viewed then as it is today. Wine was either for certain ethnic and religious groups not thought highly of then or predominantly for "winos" – alcoholics – being fueled by fortified concoctions. If you are interested you can look back at the state records on types of wines produced by year. You will be shocked at how small an amount was non-fortified as late as 1970.

Sprinkler irrigation came along in the 50's, in other industries, using aluminum pipes that had to be moved by hand. As a teenager I had to move those in the apple orchard. They were a new thing. Before that irrigation in orchards was by furrow and gravity. When Mirassou planted their Mission Ranch they used movable aluminum pipes to irrigate their young vines. Once stakes and wires were up that was no longer easy and so they heli-arced the pipes together and affixed them to the top of the grape stakes becoming the first wine grape vineyard in America to have overhead irrigation. It was a new innovation. The thinking then was that they were crazy because overhead water would cause rots and mildew on grapes – it couldn't be done without ruining the crop. Ed Mirassou was right – it could be done!

Along about this time – the mid 60's - plastic pipe that could be buried came along. This was a major gain for orchards as no labor was needed to move aluminum pipes. In early 1969 I installed the first over-tree irrigation system using plastic buried pipe in the Wenatchee Valley. I don't know about elsewhere. The local irrigation company and I went around and around about how I wanted it done – from design and calculations to installation. Again, they thought I was crazy. Like all things new there were flaws in the installation, even aspects I had warned them about, but I could not be there in person as I was a little busy preparing myself and younger pilots for a visit to Vietnam – my second. Those flaws were corrected upon my return a year later.

The buried plastic pipe became common quickly as it was so simple and cost-effective. Along about the same time drip irrigation was being developed. The main workers in this technology were the Israelis dealing with salty water and short supplies in their desert conditions. This technology was not dependably developed when the grape planting boom of the early 70's came along though some did install it then. It took about another ten years before dependable support equipment (valves, pressure regulators, etc) and quality dripline material came along. This was another major gain for grape growers.

Like all things in life, it seems like there is no free lunch. There are pluses and minuses to both systems – sprinklers and drip. Neither is cheap, at first blush, to install but over the longer haul they are very cost-effective and, believe it or not, I recommend installing *both together*! They each have significant problems that are offset by the other.

The Sprinkler System

This system is for area coverage. Its layout must take into consideration the prevailing winds, if applicable, as to direction, frequency, duration and force. If winds are a factor then the distance between sprinkler heads must be closer together perpendicular to the wind than in the direction of the wind. As in all systems design is a critical process. Today, I'm sure there are computer models to calculate all this stuff. In the early days the procedure was far more primitive but it worked. One set up a whole bunch of coffee cans or buckets around a sprinkler then ran the sprinkler for a long while in the wind. Then one simply measured the water in the cans. The distribution pattern was apparent. From that information one could then proceed with the desired layout and associated friction pressure losses.

The sprinklers have many positive aspects. They give area coverage so one is able to re-charge the entire soil reservoir allowing the vine roots to utilize the *entire* soil profile. Sprinklers allow frost protection to the limits of water supply. In areas of dusty air – as in the Salinas Valley – one can wash off the leaves and fruit. That point may seem trivial or nonsensical but heavy dust on the leaves can cause problems. It looks like fine dust to you and I but to the bad mites those specks are huge boulders to hide behind and the predator mites can't find them. Wash the leaves and the predators really go to work. I have seen vineyards with both systems where the grower chose to use only the drip and have severe mite problems. After many sprays to battle them and losing, one grower asked me what to do. I told him "turn on your sprinklers". He thought I was smoking funny tobacco. When I explained he did it. No more problems. He still uses that technique today. Further, as a before-harvest quick rinse it really helps winery pumps. Un-washed grapes after pressing can leave a several inch thick layer of fine sand in the bottom of a tank – sand that has acted like carborundum as it goes through the pumps, severely shortening their life.

The sprinklers can be used to cool a vineyard during a hot spell – evaporative cooling of fruit that otherwise might cook or of leaves that might burn, or preserve malic acid.

The sprinkler system takes a lot of power to operate at the required pressures. Today's energy costs are substantial. There is a significant water loss to evaporation off leaves. There are times one does not wish area coverage and weeds that may result there from. Wetting the vines and fruit carelessly near harvest can result in rots particularly botrytis. One cannot put fertilizers through the sprinklers during the growing season.

The Drip System

Drip systems are low pressure surface systems with emitters slowly dripping water into the selected location. The equipment is now well developed. The drip system allows more precise application of water and fertilizers at selected doses at any time thus saving tractor passes to fertilize. It does not generate high humidities in the vineyard at the time of usage. It wastes little water to evaporation. It does not limit other vineyard work appreciably. It allows fine tuning of water needs near harvest.

The drip system does have some negative aspects. It depends upon certain amounts of clay in the soil as it depends upon capillary pull to spread the water outward. In rocky/sandy soils the soil-water pattern is essentially straight down! The wetted area is very small. Some years back when drip was really coming into vogue I saw perfectly good sprinklers *torn out* and drip systems installed on just such ground. The *financial* guy was very proud of the conversion pointing out to me that he covered the cost in two years by the differential in power bills. The blocks have never in all these years yielded over three tons to the acre. There is no way to re-charge the area or flush it. The water bulb is small. So is the crop. They've been paying a lot more in lost crop.

There is no or poor frost protection with drip. Some fogger systems are in use with much added expense and I don't think they work effectively. They may even cause more damage by their cooling effect. Any air movement at all can cause the fog cloud to drift away exposing the vine to cold shock and freezing.

There is, or can be, a "salt wall" build up around the wetting area as the grower puts water in and the vine takes it out leaving salts behind. Grape roots are particularly salt intolerant and often won't cross the salt barrier thus limiting their scope to the wet ball area. If for any reason that wet zone cannot be timely recharged the vine has substantially less ability to withstand the "drought". Absent strong winter rains and sprinklers there is no way to "flush" the salt away.

The two systems used together gives us that happy world where the pluses and minuses offset each other and gives the grower tremendous opportunities to fine tune his crop far beyond anything our forebears could dream of! Using the area coverage aspect allows us to re-charge the full soil reservoir at will. This allows the vines to occupy and explore the entire soil profile. In these times of high density plantings and intensive techniques this is particularly important. We can pre-charge the soil at dormancy such that winter rainfall will purge or flush our soils of salts that may have accumulated. Of course, frost protection is useful. Cooling the vineyard will help preserve malic acid and prevent cooked flavours in the grapes and keep our vines functioning and healthy. Many observant growers have seen their vines "shut down" for perhaps two weeks or so from excessive heat. Proper and timely "cooling" precludes this. Vine washing can be used as needed.

The other system allows us to fertilize with soluble materials without tractor passes in the field. Dripping often allows us the cost savings in power and available water supplies. We can fine tune into harvest with judicious use of the drip. The post-harvest irrigation is well served by the drip system utilizing the sprinklers for area pre-charging once dormancy is entered such that the winter rains can flush.

Over time, as the grower becomes intimately knowledgeable about his soil variations, fine tuning of his systems can be done if desired. Initially a drip system's emitters need to be located very near each vine especially if the soil is rocky sand and if there is no intent to use the sprinklers often. As the vines become well established the drip should be slid over a foot or more away from the vine crown (where it emerges from the ground) to avoid constant moisture at the crown and to avoid any splashed dirt onto any wounds from suckering activity. In areas that need more water than others very small additional emitters can be easily inserted. Over-watered areas can be reduced by inserting plugs and spreading out emitters. At Ventana we changed nozzle sizes in the sprinklers as necessary according to differences in soil moisture holding capabilities.

Cognizant irrigation management is an essential tool in growing premium wine grapes. It must be used astutely in accord with changing conditions. That seems like a trivial and obvious point to assert but often I have seen the practice to be totally different specifically within larger operations. In December or January projections and budgets somehow find themselves etched in stone. They drift into the hands of bankers and chief financial officers and take on holy characteristics of scripture. No matter how much the conscientious grower emphasizes that they are simply guesses and estimates once they are in the hands of financial folks they become icons to be worshipped. Employees of the larger corporate operations learn quickly to follow the schedules and make no waves. Changing things are not in their best interests and can threaten job security. As a result we often see irrigation operations conducted in accord with a hard schedule never deviating no matter what. It is strange how often in the last thirty five years hot spells have peaked on the weekend. As it happens, *non*-owner operated vineyards and corporate vineyards are usually governed by farm management companies that have a "company" culture. They are off on weekends. Problems can occur – and do.

Grape vines do not like wet feet, as I've said. "Perched water" is water that sits on top of an underground sharp change of soil type – lenses or hardpans and such. Grapevines like drainage. Actually so do fruit trees. For that reason it is strongly urged by the schools that land prep be done before installing the vineyard. I concur. A grower gets only one shot at this very important step. Often – more than often – it is viewed casually. It is simply a matter of hiring a big crawler with shanks and having them run around the field for awhile. There is usually a schedule for the project that maps everything out on a calendar. This step should be investigated thoroughly concerning the type and structure of the soil to be ripped and the extant moisture condition. I pass on a story as a warning. Long ago, on a property north of Soledad Prison, a development to vineyard was underway and schedules "must be kept". The soil on this property was/ is heavy to clay. It had rained heavily but still the crawler and shanks were put in the field to rip it three directions crisscrossing. After smoothing a trench every 48 feet was dug by a backhoe to install the sprinkler line plastic. The dirt was pushed back in over the plastic lines. The vineyard was installed with rows every 12' thus 4 rows were associated with each sprinkler line – including the one in the sprinkler line. Time passed. The vines in the sprinkler row grew fine – very nicely. The others were all stunted or dead. Upon inspection the three way ripping activity in wet clay had polished the clay beside the shank paths through the clay making the walls ceramic-like. Thus, the field was a series of triangles with ceramic walls and grooves with ceramic sides. Vines in the triangles got little water, vines in the grooves drowned and vines in the backhoed row thrived. Ripping needs to be done in soil as dry as possible to shatter and stir it. Mixing is the goal – breaking up any *lenses* that may cause perched water.

When using sprinklers on varieties prone to botrytis one should try to make sure that the time duration from start of irrigation to finish *to complete vine* and *fruit* drying is less than 18 hours preferably 16 hours. Botrytis appears to need at least 18 hours of moisture conditions for it to sporelate and spread. Heavy dew and fog conditions must be *included* in the eighteen hour span calculations.

It is my opinion that the biggest error in the use of irrigation is the overuse of water. The old saying "If some's good, more's better" seems to be believed but in the case of irrigation (and fertilizer) it absolutely isn't correct. It is an awesome power to touch a button and have the god-like power to create rain. It is hard to resist but it must be done. *After* a vine changes from the "tendril" perception to the "tip" perception they go through another sequence. Within the normal color specific to each variety, as a vine begins to stress slightly or work harder to extract moisture from the ground the "tone" of its color *deepens*. How far into this tone deepening process one goes before irrigation will be a function of the grower's experience with his soil type and the climate. For example, vines pull water very easily from rocky sand and then, boom, there is no more while from clayish soils water depletes much more slowly and more is available. The rocky sand requires instant reaction while the clays give a little more slack if heat hits.

A grower often (usually?) finds himself in a bind at harvest time. Wineries seem to have all sorts of reasons for not accepting fruit when the grower says its ready – locked up schedules, no tanks or truck, crusher is down, grandmother has a headache, State law says no work on Sunday, etc., etc., etc. ad nauseum. As the sugars rise while one is waiting using drip irrigation judiciously will result in a 2 degree sugar drop (on properly ripened fruit – more on dehydrated) that will recover in two days or so without any water-taste in the fruit. This can be done several times. Remember the problems of 2005? Time can be bought this way. By the way, it used to be illegal to add much water at the winery but now its done. It has *never* been illegal to add water to the vines. Even if grapes are taken with a bit of water taste the magic of fermentation gets it gone. Why do you think the big wineries recently pushed so hard to legalize water additions? They know that if they have high enough sugar the water addition pre-fermentation doesn't show.

I know of one situation years and years ago where a block of Chardonnay near Gonzales was horribly botrytised and rejected. Another winery took the fruit. The water hose ran steadily as the grapes were put into the hopper and then into the press. The grapes were pressed and sugar adjusted with more water. The juice was settled, racked and fermentation initiated. During fermentation some Polyklar AT was used. The wine turned out lovely and later on received many gold medals. I rest my case on this observation.

The Devil is in the details.
—Saying

To choose, it is first necessary to know.
—Herman Finer

Section 3H
PRUNING

Fruit plants – vines and trees – left to their own devices will produce short bushy shoots and small fruits. It is sufficient to achieve their needs but not ours. Long ago humans discovered that cutting back the previous year's growth, while dormant, to just a small portion of the total buds grown resulted in much stronger growth and more desirable fruit. From that time forward pruning has been the subject of arguments, discussions and studies among growers. It will probably go on forever. How much to cut off and how much to keep is always a tug of war inside the grower. Long ago my stepfather and a neighbor for years traded orchards when it was time to prune. I asked why, they laughed and explained to me that a grower just couldn't take off as much growth as he should on his own trees but had no problem doing so on someone else's trees. It worked for them.

Pruning within a given vineyard system is pretty straight forward once certain principles are understood clearly and religiously conformed to. If the vine is thought of as a pressure "irrigation" system providing liquid and nutrients to growing points then some conclusions are obvious. Also keep in mind the basic function of the trellis. It is expensive to install and to maintain. Every running inch of that bottom wire – and some of the second wire – is the production line of the vineyard. If it is not used *completely* then one is running below factory potential. One severe problem we face is that our accounting systems do not measure "potential". That is, *what should have been.* They only measure what was. Thus, there are no definitive flags waved until the shortfalls fall below the lines of red and black. Usually even then the short-ages are laid off to "that hot spell in June" or "those rains in July" or to some other mystical non-measurable event such that no one is responsible except Mother Nature.

When I was designing this system in the mid seventies there were many criteria that needed to be incorporated. The training and pruning of the vine, within the fruit quality limits, needed to be handled. Another factor was to have the design be ergonomically correct for the work force. The California haystack system was difficult to prune. "Difficult" means expensive. It also means that as the day progress-es the people became very fatigued. A fatigued workforce means expensive as productivity decreases. One wants the work area directly in front of people so there is minimal reaching and muscles work the way they were designed – both for handpicking and pruning. We also needed to de-sign such that that production line is operating as close to 100% of potential as we can reasonably get. In some sense everything else we do is fixed in that one sprays the entire row, one runs the harvester down the entire row, one discs or mows the entire center, one does weed work in the entire row, etc. To the extent that crop is below what could be, the per unit cost of production goes up. "Goes up" is not a good thing – or so my Grandpappy told me! Another thing we had to consider is keeping air movement across the floor of the vineyard for humidity control. Here in Botrytis country that, wind and weed control were some of the reasons I rejected considering the Geneva Double Curtain or any other "hang down" system.

Faithfulness to the past can be a kind of death
above ground.
—Jessamyn West

Section 31
TRAINING

Recall earlier that I spoke of adjusting the vine spacing within the row according to the expected vigor and capabilities of the soil, root and climate complex. Many of our varieties on the Ventana were of relatively low vigor on their own roots in its rocky low-nutrition soil. As such, the 3 ½ foot spacing between vines was very appropriate. On the more vigorous varieties such as Sauvignon Blanc (talk about a redundant name!) or Syrah we would increase that distance adjusting it according to soil characteristics the row was passing through. These distances would be as much as 4 ½ to 5 feet or maybe, here and there, a little more. It is my opinion that at 4 feet, and definitely at 5, the "yoke" system (Bilateral Bidirectional) should be used to divide and balance the liquid flow from the roots. Long runs on the cordon arm in cool climates are just too difficult to manage and detrimental to crop uniformity. I chose to design to a unilateral, unidirectional method for most of our varieties. This system allowed the bending and tying of the growing young plant to a nice tight curve as it approached the bottom wire. That bend area is often called "the shoulder" and that is a critical area in the long run. The shoulder MUST be *flat* as it becomes a cordon! If it is not flat but rather bows up and comes down to the wire it creates a high point. Apical bud dominance begins to work and it gets first call on the sap. This will create "Bull" canes in that area and cause delay and/or stunting further out the cordon. (SEE PLATE 19).

Further, consider that the "pump" in the vine must push all the sap to higher altitude before it comes back down to the cordon level. That is unnecessary "work" or energy expenditure. Keeping that area vacant will take time and attention throughout the years. This curvature commentary may seem picayunish but drive along and see how many vines have this feature in existing vineyards. Everybody knows it but failure to have tight supervision and training on this aspect is the cause. One effect of this problem is to waste space on your cordon wire as one must *never* allow growth in this area. Never. Did I mention never? The "wasted space" loss of the yoke can be regained by bringing a cane back over that area on the second wire – but do not use one from the first spur.

Another aspect of "bending" the growing young plant to the wire is that forcing the shoot to grow in the horizontal slows the growth a bit and during the period of elongation of the internode cells they do *not* elongate as much as they would if allowed to grow upright. Those buds on that shoot (soon to become a cordon) are one's future source of meristomatic tissue or spur sites. In my mind the more of them the better within limits. I have seen many vineyards where the shoot was grown vertical to be tied out the wire the following winter, grown in tall tubes, fertilized like crazy or all of the above. The internode lengths can be 8 inches or so. In warmer less windy areas they can even be more. If one wishes to have spur sites at 3, 4 or 5 inches apart you've shot yourself in the foot before you ever start! It is far easier later to remove a bud than it is to graft one on! I've been in vineyards where the manager was quite proud of how great the vines were growing as above but became very concerned when I asked his desired spur spacing and how he was going to accomplish that. So often in this business its one job at a time without considering the ultimate goals and down-the-road implications.

The tieing of the growing shoot out the wire continues until the *last* bud on a given vine is at the desired spacing location before the *first intended spur* bud on the next vine. SEE PLATE 14. Not before the shoulder bend of the next vine but before next *intended spur site!* It emphatically is not the shoulder. One should not allow any growth in the curve area. Once the cordon is completely horizontal *then* consider the first spur site. In this fashion one should theoretically have a continuous function of distribution from end post to end post. By keeping all growth off the curve one will achieve relatively uniform distribution of the plant's liquids.

In the case of "yoking" the vine the historical procedures are fine except that considerable time of development will be lost. As I mentioned earlier, in the years before growth tubes the heading off would cost a year in Monterey's cold windy climate. Now, with growth tubes that isn't so but still it significantly delays full cordon development. The idea of not using growth tubes because of their full cost (tubes, installation, removal and disposal) is a false economy when one considers one year's interest and carrying cost of the project. The earlier one can head-off the vine the better in my opinion. If the vine growth is such that one does not have sufficient to head-off by mid-June to early July one should probably not head-off until winter.

The "yoke" should be formed as tightly as possible and the curve dis-budded during the following winter's pruning. Dis-budding the tight green curve can result in the shoot snapping even if already tied because of stress in that area. Do not worry a great deal about the gap on the wire in the yoke area though it should be tight as possible. We will pick that up later by using the second wire. Again, the cordon arms should be cut such that the last bud on one arm is exactly at the desired spur spacing from the last bud of the next arm. One is trying to have all the wire occupied by properly placed spurs EXCEPT any in the yoke area. It is still *imperative* to not allow any shoot growth in the yoke area. When inspecting the work *while* it is underway one must keep repeating "Nada in la curva"! The supervisors must be taught to emphasize the point. Following this simple rule and enforcing it will result in more uniformity of shoot growth the full length of the cordon, more uniform bloom and fruit with a tighter deviation around the mean sugar degree. As I drive around I see many vineyards not following this method. There will be problems ahead removing those eventually and they are suffering now but don't know it.

In year 2000 I bought a nearby vineyard in which this rule had not been followed for twenty years. There were massive spurs points in the yokes. On the cordon arms the cane growth tapered down at about a forty-five degree slope – very long canes in the shoulder and yoke areas, twelve inch shoots at the end. In those twenty years that vineyard had never exceeded 2 ½ tons, some years less, until 1999 when it yielded 4 tons, harvested at the end of December and did not make sugar. That winter it took small chain-saws to cut out that massive growth in the yokes and off the shoulders. The vineyard yielded the next year in excess of 8 tons at full sugar in early October – on time. Admittedly, there were other farming factors at work. The shoot growth was generally uniform but it did take one more growing year before fully acceptable uniformity was achieved – primarily to build up vine stored reserves. They had been depleted in 1999 completely.

The year after the cordon cane has been laid down the vine may/will try to overcrop itself. It still does not have the rooting system or stored reserves to carry the crop that it will try to carry. One should realize that one is still in the "timber business" – growing wood. Dropping fruit clusters very early should be budgeted. They're still teenagers – don't overwork them.

The time for spurring is straight forward – the vine will tell you if it wants 1 bud or 2 bud spurs. All one has to do is listen. If the vine gave you the desired amount of cane growth the preceding year – 2 bud. If not – 1 bud.

The next phase is a little more problematic especially for growers who use farm labor contractors. The problem here is teaching supervisors so they can teach and monitor the pruners. Recall that I mentioned that in the old system the Chardonnay could handle around five feet of fruit wood yet I chose to plant at 3 ½ feet.

This system not only wants continuity along the cordon but it also is set up to use the vertical plane to balance the vine (again, the vine will tell you) and distribute the hanging fruit such that we avoid fruit to fruit contact as much as possible while allowing light to reach the fruit. The old haystack system packed the fruit close together in the dark tunnel down the middle where the sun didn't shine and the air didn't move. If we observe a young shoot we can see that Mother Nature has staggered the clusters on the cane – one above the other at some distance. This system follows the same process by putting a "kicker cane" either up on the second wire or up, along and down from the second wire. This accomplishes several things. It allows us to increase crop to the load the individual vine wants to carry. If it's weak – spurs. If it is a bit stronger – a "boot jack" to the 2nd wire. Stronger yet – up to the second wire and a wrap. Strong – a cane along the second wire. Really strong – multiple canes. Actually, trained pruners *can* make these decisions if supervision is tight or if they are employees of long standing with an operation. This "balancing" of the vine promotes uniformity in the fruit at harvest and is conducive to the long-term health of the vine. Further, by going "vertical" we introduce fruit separation by up to 12 inches making a "wall of fruit" of one cluster thickness.

Of course, this method of pruning with its associated decision making and tying of canes is more expensive per acre than simple spur pruning even with mechanized pre-pruning. However, no matter how much the reduced cost of straight spur pruning may dazzle the accountants and bankers because of myopia there are severe drawbacks associated therewith.

Within our industry (and among the bankers servicing it) there is a hard fixation upon accounting by job function and by quarter. Again, this fixation leads to irrational thought processes from a systems point of view. If two growers submit budgets and one grower has a significantly higher dollars per acre allocated to a job category than the other the banker and/or the CFO goes bonkers. In the early 80's and on it was a serious problem because the banker simply couldn't understand the explanations and, more importantly, couldn't explain them to his loan committee. The concepts were too new.

One of those concepts was the crop-limiting aspect of spur pruning. That is, the *upper limit* of crop size is set at time of pruning while the down side is unknown and the growing conditions of the year are also unknown. To understand this we have to consider fruit-bud differentiation – when and how it occurs. As recently as one year ago I saw a newsletter discussing a bountiful crop because the spring weather was so warm and good that many clusters were formed. Let me be perfectly clear – fruit clusters are NOT formed in the year of harvest! They are formed within the bud the growing year BEFORE. At that stage they are called "initials" and can be observed within the bud by slicing the bud with a razor and, using a glass, can be observed coiled up therein. In 1980 we formulated this technique using it *prior* to pruning to get a "feel" for fruitfulness of the basal buds and others and adjusting the amount of wood left accordingly. This technique will give the grower some indication of the potential clusters per bud by location.

What it will not do is tell the grower the absolute size of the cluster nor the viability of ovules or pollen. That remains to be seen and is a function of nutritional, climatological and energy conditions. The size of clusters can be affected by stored energy within the vine from the year before. Shortness of energy can cause reduced growth across the vine.

Cold weather will cause less photosynthesis substrate being generated than warm, etc. Shortages on nutrients (such as zinc and boron) can cause ovule non-fertility.

Thus, straight spur pruning sets a limit before the chess board of the current year becomes clear to the grower. If the preceding year's conditions were such that at the time of fruit-bud differentiation only one fruit cluster initial was formed per bud and it was a tiny cluster then the crop load this crop year will be severely reduced from the "potential" yield – even if this year is a gorgeous growing year. It is entirely possible in cold regions that basal buds on many varieties can be non-fruitful if the conditions at differentiation time were particularly onerous and of long standing. Typically, buds further out the cane undergo differentiation later on in the season and thus contain the more normal two-initials per bud format. Often we see, for example, one-initial or blank basal formation in Chardonnay while seldom in Riesling. Riesling, however, bud-breaks later in the spring than Chardonnay (if pruned at the same time) and thus forms in a little different (perhaps warmer) time-frame.

Some people view Mother Nature as functioning by an ever evolving grand plan or following a plan of Intellectual Design. Both of which imply a "plan" or "order". I am of the opinion that Mother operates in a mindset of "chaos". My take on it is that MN is constantly taking a mix of factors or genes or whatever and throwing them into the "pool" of Earth and waiting to see which "swims". We observe this all the time in all its cruelty and success. It is not the "plan" which is genius – it is the "process". That "process" itself is the factor that gives us successful adaptation of organisms to our ever-changing environment both micro and macro.

It seems to me the organization of the grape cane is no accident of design but rather a much needed organization to accomplish its survival goals. Look carefully at the location of each of its parts. At bud burst in the spring the new shoot begins to grow. The initial leaf precursors were already curled up inside the bud. As the bud grows and extends the initial leaves unfold and begin their growth. This growth is entirely a function of stored energy. In their youth leaves are energy "users" and not until later are they "generators" and contributors to the vine. If one looks carefully you will see that the leaf attachment to the shoot is the very first tapping into the sap flow of the shoot. That is, that new leaf gets first call upon the stored energy in the flowing sap. Immediately *next up-stream* and adjacent to that leaf attachment point is the forming new bud. Why exactly here? Why not the other side or elsewhere? As the leaf matures and becomes an *exporter* of CH to the vine the very first organ in the line of flow is that bud tapping off its needs the remainder allowed to continue up the xylem stream to whatever may need it, other developing leaves, buds, tips, etc. Why is this important, we may ask?

Like the *fruit* in a given year preserving, by seeds, the species, so the bud is the preserving entity for the *next* year. Now, the perennial plant doesn't have to be successful every single year. It has a long life and it can produce many seeds in a given year, or skip a year or produce more or less. But, to accomplish its longer term program it must survive the short term. Now, we humans/winegrowers being under rules of economics need crop every year. But, what we observe is the bigger crop, lesser crop syndrome. I've covered that in the energy discussion however it is in the bud and its location that this mechanism is expressed.

In making its decision to produce in a given next year or to pause entirely or to reduce its seeding the bud must have some criteria upon which to base its decision. I believe that that *decision* is based upon NUTRITIONAL conditions *NOT hormonal*. Here I want to be clear that the powerful hormonal controls of the seeds can negate the advantageous positional aspect of the bud but the *bud's decisions* are *nutritionally determined then* hormonally controlled.

Fruit-bud differentiation – that is, the decision to make a future-shoot either fruit bearing, non-fruit bearing or partial fruit bearing – does not occur immediately upon bud formation nor is it made while gestational. If the decision making was programmed early it would fail because the vine and leaf is making such a demand upon stored energy but more importantly the vine would not yet have had a chance to "sense" the present season upon which its decision must be based. Ignore the folks who speak of present year fruitfulness because of a "warm spring". It is nonsense. The dormant bud made up its mind the previous year. And the decision was sequential out the cane at different times throughout the growing season according to the conditions extant at each bud's decision time. The future *number* of clusters and the *number* of *potential* berries was determined at differentiation time. The number of berry seeds that set, the size of berries and length of cluster from within the confines of the variety are a function of the seasonal growing conditions and stored energy of the vine. Ovule viability and "set" is a function of both nutritional conditions and weather conditions at bloom time.

Long ago Professor Jim Cook of U.C. Davis gave a nutritional talk to we Monterey growers – sometime in the mid seventies. He put a chart up on the wall showing nutritional pick up by the vine. High up on the chart was a wavy line showing soil applied additives. A little line way at the bottom displayed the data they had experienced with "foliage" applications. He said because it was so minor foliage application studies had been dropped from the work and were no longer being done! It was that little line that grabbed my attention – it was positive! That little line meant we could "feed" the vine at *critical* times. Annual feeds – no, at least not economically. But marginal "boosts" at critical times – YES, with capital letters!

Besides "set" needs of zinc and boron (if needed) foliage feeds *prior* to bloom, certain weather conditions or reduced stored carbohydrates may create a situation where foliage feeding of a "kitchen sink" may be beneficial for fruit bud differentiation of lower buds. Obviously, removing mature basal leaves before differentiation reduces CH flow to the bud. On cold and windy springtime Monterey post set I have often used this approach in a "flood the zone" goal. In these situations it is also beneficial in lengthening the cluster frame during its period of elongation.

Another observation and sequential line of contemplation lends support, I think, to my contention that fruit-bud differentiation is nutritionally determined. Keep in mind that academia (see General Viticulture, Winkler et. Al, 1964 and other writings) has consistently directed that when pruning a "sun cane" should be kept – not a "shade" cane. The sun-cane will have a different color when dormant – and that is correct. The assertion was that the shade cane's buds were far less fruitful which led to the general idea that sunlight causes fruit formation. This instruction was particularly important then as the California Sprawl or Haystack system generated shade canes. Most growers would agree with Professor Winkler. I, too, agree with the procedure but do not subscribe to the idea of the sun being directly the agent rather the causative agent indirectly (and Winkler made no such statement). I think the mechanism at work is again one of species survival and spread of progeny.

Consider the color of the young growing shoots – they are green in all their parts – newly formed buds included. Not just the leaves photosynthesize – all green tissues do so – it is the chloroplasts that give the green color. Their presence means photosynthesis if there is access to light. The skin of the new shoot and the "shell" around the new bud are generating CH. The effects are *additive* to total CH available but the CH generated at the bud is localized on top of that being supplied by the vine and from the newly matured associated leaf. That *marginal* difference may be critical in the decision process.

Shade canes are determined by color of wood – not by diameter of wood necessarily. That would indicate that early in the growing season the growing shoots in question were all receiving sufficient *stored* CH for their development. At the proper time lower leaves were contributing normally. As the mass of shoots and leaves developed rapidly the shade cane's environment changed and no light was reaching the buds and thus no CH production occurring there. The associated leaves were also being shaded out. Yet the supply of stored energy being transmitted by the vine in the nutrient flow past these buds is the same as for the sun cane. The sun cane's buds are receiving "energy packets" and are thus at a higher energy level or warmer which could contribute to the process. It is that marginal differential that the vine uses in its decision making process. One might note that the new trellis system takes this into account and generates no "shade" canes.

In cool climates there is usually an initial spring period of some growth but the pattern is of really cold periods and other lousy situations until the weather gets its act together. The winds in Monterey can be quite heavy during the period *before* there is enough shoot growth for putting up trap wires and mechanically elevating the wind thus increasing temperature in the foliage wall. It is *precisely* during that period that the earliest-formed buds are beginning to make their decisions. Cold and windy weather have the effect of reducing newly generated CH and transference of stored CH. The *rate* of vine processes is a function of temperature peaking at 85-90 degrees F and declining thereafter. This spring/early summer situation can lead to a reduction in fruitfulness of the lower or basal buds when compared to buds further out the cane which differentiated in warmer or less windy conditions. Thus, when one spur prunes one's cluster and berry count in year two is a function of a twenty day weather condition of the previous year one! Once that bud has decided – it has decided and there is no changing that.

Later, as the green tissue has served its purpose, the vine then commences its protective work for the coming winter. It hardens that "shell" around the bud with dry material to give the live material within protection – dry material that can absorb winter moisture, expand and weaken the previously sealed unions. It also abandons the green skin material of the shoots and commences lignification of the shoots beginning at the earliest formed end (basal) and working outward as time allows. At the end of a growing season one sees small green portions of the cane end if shearing has not removed them. The "ripening" of the shoots into canes at and after veraison is of utmost importance. This work is being done in conjunction with the development of ripe compounds in the fruit and providing storage of CH for the next year's crop.

It would seem to me that if I were interested in spreading my children as far away from the competitive area around me I would want to put my berries out in the open as far away as I could so the animals could get to them – not buried inside a mass of foliage and wood. If I were a "looking for sunlight" plant I would also want to have viable buds away from my own shade. It seems to me to be sensible to form protective sites early (one doesn't know what the rest of a temperate climate will bring) then focus on extension as the season allows.

Some varieties such as Thompsons are notorious for no fruit in the basal buds – these varieties don't even bother wasting energy there. Many, many of our varietal wine grapes show the same tendencies. Rieslings, for just one example, will often display one or two clusters from basal buds but then give threes from buds further out the cane if the preceding year was generous.

One may say "so what" on all this. Grapes are the vehicle from which both vineyards and wineries earn their income. Given that, it sure seems appropriate for a *grape* grower to understand how and when a vine decides to give us the grapes. Then, the question of intervention arises in order to accentuate the positive and eliminate the negative – as the old song goes. Let me give an example. Years ago, Gewürztraminer was notorious for alternate bearing – 5 tons one year, zip the next for a statewide *average* of 2 ½ to 3 tons per acre (U.C. Davis yield publication for varieties). This sounded very familiar to me because Golden Delicious apples had that same habit until growers learned to break that cycle with specific nutrients at certain times *and* to not be greedy in heavy set years. Applying that to Gewürztraminer we broke the cycle and stabilized annual yields. One year along the way I decided to skip foliage nutrient applications – because I had no money not because I wanted to. The next year's crop was down dramatically! I tried in the future to not let a little thing like no money stop me – wasn't always successful at that trick, though.

I submit the idea that farming procedures can affect the regularity of bud fruitfulness, size of cluster count, size of cluster frame and potential berry counts and berry size at harvest (big if desired, small if desired). These variations can easily cause ranges of 3 to 5 tons per acre! You doubt? A nearby chardonnay vineyard in 2005 (big crop year) yielded *13* tons to the acre. The year before was 7, the year after was 6. Much of the big crop/little crop, hot year, cold year explanations usually stated are really a function of not understanding the above discussion in all its ramifications.

If straight spur pruning has resulted in the proper vine load over the years then the vines were planted too far apart in the row, the vertical element of the trellis has not been used and the *land* is producing below its potential. In the mature vineyard suckering is a variable according to vine count. Pruning is a function of linear feet. Most of the other factors are "fixed" in the sense that one does the same job per acre not as a function of "vine" but of "row". A reduction by half of crop load means that one's cost per unit of production doubles. To use an example in absolute dollars if a yield of 4 tons is achieved when the potential was 8 tons then, at $1,000 per ton, a $4,000 shortfall per acre would be the result of designing for a $200 per acre pruning saving differential. Our accounting system's failure to recognize "potential" and flag it leads to this sort of thinking. A "by the job" viewpoint contributes to this failure to recognize the loss of potential.

If necessary, I have found that it is far easier (though expensive) to green drop crop than it is to graft clusters on to the vine. If you do graft you may upset your neighbors when they find out where you got those clusters you are grafting on. One may note that green dropping is done AFTER the grower sees the chessboard of conditions prevailing in that growing year thus allowing the fullest grape yield consistent with the finest quality. By then it is a "choice" for the grower not a "dictate" resulting from a decision made six months prior or at the time of planting.

I have noted elsewhere that too light a crop is detrimental to wine quality. It absolutely is detrimental to the health of the grower.

Varieties differ in their time of bud push in the spring. If there is a warm spell many will push similarly but in cooler weather there will be distinct differences which observation seems to give credence to the pressure/temperature line of thought. For instance, Chardonnay likes to push early and then just sit there for awhile all exposed to potential frost while Cabernet Sauvignon delays its push to several weeks after the Chardonnay.

The "time" of push will vary *within* a variety according to *time* of pruning. Early pruning will lead to early push. This feature can be used by the grower, if desired, to control the rate of grape flow to the winery. While many (most-all) big wineries want the grapes of a variety to flow from a grower at nearly the same time frame smaller wineries have flow-rate limitations particularly as the season moves into its later stages. Staggering pruning times within a variety will stagger the ripening process so that everything isn't ready at once. It is another tool in the arsenal for a grower to be used if needed.

I have seen Chardonnay push in February then sit exposed to potential frost for two months! When buds start to swell from an early warm spell an overhead sprinkler two or three hour run during the warm part of the day will substantially delay the push by cooling the vine! I have done this long ago but couldn't cover all my Chardonnay. The un-sprinkled vines pushed. The sprinkled vines delayed a month plus. The weather turned colder after the heat spell. The cooling appeared to reduce the vine internal pressure. I also suspect that the film of water on the vine absorbed the light quanta thus interfering with their penetration into the vine. The same can be said for the water on the ground not allowing penetration of quanta into the soil precluding soil warming.

In frosty areas delaying pruning can delay bud push hopefully beyond the frost season. One old-time trick was to do the gross pruning at the regular time leaving 10"- 12" long canes at each to-be retained spur. When the buds push – and it will be the upper buds that push – the grower could *very* quickly go back through the vineyard cutting those 10" canes back to two bud spurs. This will "shock" the vine a bit by pressure reduction and delay the pushing of the spur's buds maybe two weeks or so and, hopefully, beyond the frost season. I know – I'm repeating myself but this is on pruning.

When selecting a cane to be placed on or over the second wire it should be sourced from a different spur site each year. It should never be sourced from the same spur site as the previous year! If done this will excite the apical bud dominance trait and lead to enhancement of that site and decline of the remainder of the cordon arm. It should never be sourced from the first spur site after the bend or yoke. A classical error I have seen repeatedly done still today is, in the desire to produce in the early years, is to keep a cane from that first spur-site up and back over the curve area. Usually that one is kept because it is the biggest or, often, the only one keepable. When that occurs (usually) the vine is telling you it needs another year of development. That spur site sizewise, becomes dominant and a cane from there is used each year. SEE PLATES 18, 20 and 21. The remainder of the cordon declines in vigor remarkably. At this point pruners then become "locked in" as the only usable canes are from there – or so they think. To correct this situation often part of a year's crop must be given up to again balance the vine. That site must be reduced to one bud only and during the early growing season it will try to grow a "bull cane" – fast growing and big. It must have its tip broken off early and maybe again midseason. This will force the growth to be out the cordon. This reversal of error can usually be done in one season but sometimes it takes two. Once commenced one *must* control that bull cane. It is of very poor quality wood and will rob the rest of the vine substantially to the detriment of the next year if not tipped.

Some varieties have a tendency to push early buds on a cane and the last bud. Riesling has this characteristic and Cabernet Sauvignon also though to a lesser degree. I have walked Riesling vineyards in Alsace and Germany with growers snapping off tips on canes growing from these locations in the spring (June). Initially, until I got the hang of it, I would periodically snap a wrong shoot and would be called a "saboteur" by my grower friend!

The "Platonic Ideal" of pruning would be that each of two shoots per spur, from all spurs on a cordon and from each bud on a retained cane would grow at equal rates giving about 4" internodes. The 1 ½ to 2 average clusters per shoot would then generate sufficient load to pull the shoot tips to a stop at about 16 to 18 total leaves thirty to forty days prior to harvest uniformly across the entire block. Boy, would that be nice if achievable!

It's not. But it is the goal to which we should strive. With that "Ideal" in mind corrections to pruning and decisions at pruning would pretty much be obvious over time.

Section 3J
SPUR SITE CONTROL AND RENEWAL

During the trellis installation phase the grower is deeply occupied with supervising the details among which are the heights and spacing of wires. The tape measure is constantly being used and the workers are apprehensive every time the boss whips out the tape to double check the work. For some strange reason the critical importance of this at installation time loses its importance in the mind of many managers as time passes! That wire spacing is there for a reason. Failure to manage spur site heights completely defeats that purpose.

In the past there were beliefs about spur site difficulties in regeneration (on certain varieties) that simply were not the case – Chardonnay being a case in point. Other varieties such as Chenin and Sauvignon Blanc generate so many adventitious shoots on the cordon arm that shoot removal is often practiced. Caution is advised because those adventitious shoots can/could become needed as spur replacements.

As the spurs grow vertically with each passing year the distance between them and the second wire becomes less and less. SEE PLATES 16, 17 and 18 for a pictorial view of a vineyard wherein the spurs are *above* the second wire. This picture was taken in February 2008! There used to be many examples of this phenomenon but fewer now. In recent years the remedial work had been to choose an area and forego much crop for a year by a massive slashing of spurs which has also left many blanks on cordon arms by faulty methodology which removed the spurs tight to the cordon arm.

An ongoing spur maintenance program should be part of the instructions to the pruners. At the point in age of the vine that the spurs achieve 4 to 5 inches in height pruners should be instructed to remove at least one and no more than two spurs per vine per year. The spurs should be cut about one half to three quarters of an inch above the cordon arm to allow for "drying" back. This will preserve those latent bud cells at the base and they will push the coming spring. These larger cuts can make one susceptible to Eutypa if one is in a Eutypa area – as most of the coastal area is especially if Madrone trees are present nearby. Using a water-based latex white paint mixed with zinc and a fungicide (and IBA if desired – Indola Butric Acid – a growth stimulator often used on cuttings) paint the large cuts. This actually goes quite quickly and one person can keep up with a very large pruning crew. While at first it may seem costly it really isn't. The alternative can be very expensive.

In the past the original mechanical harvesters were very abusive to the vine often breaking spurs out of the cordon. The new heads are more gentle. However, when faced with gaps in the cordon arm regeneration can often be caused by making a slightly angled longitudinal slice with the blade of the pruning shear (whether dormant or growing) across the missing spur site and rubbing into it a paste made of water and IBA (IBA comes as a powder, the IBA mixed with talc). Usually, the gaps are readily apparent to the irrigator traveling on his ATV up and down rows.

Sometimes the situation is so onerous from past failure to manage the situation that the best solution is to cut the arm off completely and retrain a new cordon from a new shoot coming from the trunk. In the mid to late eighties when we were converting the Ventana entirely to the new system part of the conversion was lowering the cordon height from 40 inches to 28 inches and this obviously involved cutting the trunk off and training a new cordon. It was an amazing how much stored energy was in the cordon arm! One would have thought that the new shoots would have had outrageous internode length but that really wasn't the case. However, we did tip or head one of the shoots – allowing the others to absorb energy – and used a side shoot from the headed shoot for the new cordon arm. Thus the internode length would be proper for our new cordon arm. That winter the other shoots were tightly pruned off and the bow of the new arm disbudded. When the trunks were cut painting that large a wound was done for protection. Those vines exist today.

There are many ways to accomplish spur site maintenance but the most important aspect is manager *awareness* of its need. A continual annual approach will eliminate down-the-road large problems.

Another repetitive point for managers to insist upon is that when selecting a cane it should *always* be *above* a retained spur. Too often I see the reverse even from my own crew from time to time.

Pruning when the vine is bleeding in the spring does no harm that I have observed although my guess would be that its not for the best. I can't believe that the vine is pumping all that fluid without nutrient for no reason at all. I have been forced to prune a block *after* a lot of buds pushed and it was severely detrimental to growth and vine development that year and the next. Sometimes Mother just puts you in a bind and you have to live with it but pruning after push is not good except for the above-mentioned frost delay technique.

One can resist the invasion of Armies;
one cannot resist the invasion of ideas.
—Victor Hugo

Section 3K
DESIGN

In the design of this system many factors came into play. One major one was the winds of Monterey. In times past the farmers of Monterey lined across the valley, perpendicular to the wind, rows and rows of Eucalyptus trees when those types of windbreaks were thought to be effective. They were really quite ineffective and have since been mostly removed. They were trashy, used water and occupied an inordinate amount of otherwise usable cropland. Their control of wind was minimal, if any, and often exacerbated the problem of wind velocity at downwind crop level. In the 1930's, 40's and 50's there was enough paper work on this subject by the nations' land grant colleges to sink several aircraft carriers. Even today one can see, here and there, modern plantings of rows of windbreak trees which are ineffectual and expensive. At the lower part of the trees the wind can actually be a greater velocity than the unimpeded wind! If one observes carefully one can see that the downwind first and second row of grapevines will often be stunted both from wind and from root competition from the trees. This was not the solution to the problem.

As mentioned previously, the wind would pile up the shoots one over the other to the south side of the trellis. Of more import was the "stripping" of the boundary layer of air from the leaves causing the stomati to close and essentially stop the gas exchange with the atmosphere. Even with sunlight and warmth that stoppage causes a shutdown of photosynthesis substrate production as the carbon portion comes from the CO_2 in the air. Actually this knowledge was of longstanding and well noted in those plant physiology textbooks I mentioned. By 1975 the need to solve this problem was readily apparent and was continually in my mind. A solution was in hand by 1978 as the entire system was coming together.

Orienting the rows parallel to the wind flow is not advised because it creates a wind tunnel situation and the stripping of the boundary layer is increased – both sides of the leaf wall being air stripped.

The solution was to lift the wind and suspend it above the vineyard. Consider Diagram F – the rough diagram of the existing Davis system in wind conditions. The shoots came up and over the second wire – if "shoot positioned" or the cordon arm simply rolled over at point y (sometimes snapping) if not. The wind rolled over the top and impacted the next row down low at the cordon height. Notice it also impacted the "back side" of the canes and foliage and through the dotted line. It had a full twelve feet of downhill running distance. Without changing anything but row spacing look at the effect. The vertical dotted line is a "pretend" row at 6 feet: The notation "x" is the distance one could elevate the wind impact point thus raising the wind. This was still not good enough. The vortices associated with this method would still debilitate the stomati. Contemplating that sketch in Diagram F will lead one to the sketch in Diagram G. Here we have lowered the cordon wire, grown the shoots straight up and mechanically held them there with movable wires. The wind is thereby lifted and held above the vineyard by the wires. Heading or tip shearing causes the canes to harden or – if pulled to a stop – tips flopping around a bit which hurts nothing and can help stop their growth. The "run" of the wind is short and it is forced up and over by the physical restraint. The sections noted "A" and "B" are nearly wind-stripping free except in extreme wind conditions which are very unusual and of short duration. Most of the leaves were

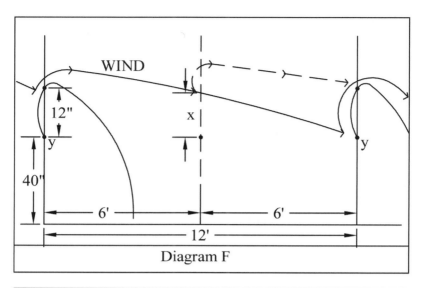

Diagram F

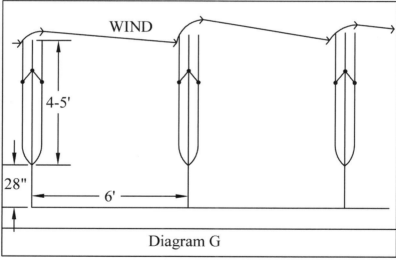

Diagram G

then able to manufacture CH (carbohydrates) throughout the usable sunlight hours without wind interference or stoppage. Water loss was dramatically reduced and the air temperature in the zone "A" was increased because the cold air blowing off Monterey Bay was not getting into that zone. This theoretical work was in place by 1978 and in 1978 the first rows in this format were planted but, of course, not yet fully implemented because the vines were babies – the upper portions were not yet needed.

When contemplating the "strength" requirements of the trellis no particular engineering calculations were necessary. The Davis pattern was so universal that all that was needed was observation. The method then was one stake per vine at every seven feet. There were many warm areas where the vine did completely fill the trellis in the "haystack" fashion and, every so often, would experience very high winds with a frontal passage. I observed that their trellis did not topple over thus one stake every seven feet seemed of sufficient strength to hold seven feet of foliage. High math! Thus, when we inter-planted at 3 ½ feet we did *not* use a regular stake. Quiedan was able to find for me some short bamboo sticks that we stuck in the ground by the young plants and used "twist-ems" – paper covered wire like you see on vegetables in the supermarket – to tie the bamboo rods to the wire for training the vine to the wire. In our sandy soil the bamboo would last for years. As a curious aside, there were other growers who watched what we did and, without speaking with me, copied the use of bamboo but on soil heavy to clay. The bamboo rotted quickly and those growers were then MAD AT ME! Had they asked I would have warned them of this tendency of bamboo. Later we found that the bamboo did have to be picked up out of the vineyard because it would come free and jam a mechanical harvester. We switched to metal rods but still it is best to eventually remove them but not absolutely necessary.

Eventually we found that as we space vines differently in the row we could shift our thinking from "stake at the vine" to spacing stakes and "Q Posts" (QUIEDAN TRADEMARKED posts) as needed and use rods for training. The modern harvesters – at least the Braud – are so gentle to the vine that the stake support isn't needed. The old "elephant ears" drug around the vine and stake while the new collectors are "traveling" and have no dragging effects upon the vine stake or rod.

By lowering the cordon wire from the Davis "high wire" system we further strengthened the trellis. Originally the high wire system was needed because the elephant ear carriage unit on the early harvesters was massive and it needed clearance. Also, the "hang-down" aspect of that system required, or so it was thought, distance to have leaves. In California thought bigger *was* better! But harvesters changed eventually. The big aspect of the wire height became worker ergonomics.

We intended to go vertical. We also needed to maintain air movement and ventilation along the vineyard floor that "hang down" interferes with. That air flow was necessary for dispersion of humidity that could excite botrytis and for efficient drying in the event of late season rains and even irrigation. In general, our workers are not tall people and the "work centers" should be structured to accommodate their size if possible. I believe the work area should be near waist high. The 28" area seems to be the best compromise. This allows for adequate air flow under the vines. For hand-picking of fruit that puts the fruit in the waist to armpit area. For pruning the bigger cuts are made at waist level allowing shear usage in accord with human muscle structure. Shear usage is thus with the least reaching. Moveable wires hook-ups and de-hooking are done at about shoulder height or a little above. This is a big thing. Try working all day with your arms up over your head! Fatigued workers are inefficient workers – work productivity plummets as the day goes on.

The work must be designed to be compatible with human structure as much as can be! Poor design leads to inherent hidden costs that management isn't even aware of because they don't do the actual work. The overly fatigued workers are doing their best but tiredness means higher labor cost per unit of production. It also leads to more accidents. If the work is contrary to human muscle organization it will lead to more doctor visits by workers and even long-term debilitation.

By placing the cordon wire at 28", the second wire 12" higher and a third wire 12" above that one has mechanical support at about the 4 ½ foot height. If one wishes higher support from trap wires they can be looped over the top of the grape stake or connectors can be attached to the upper end of the stake. However, the 4 ½ to 5 foot area appears to be sufficient. The canes themselves have a certain amount of rigidity and will extend themselves upright another 18" to 24" particularly if sheared or cane tipped. Earlier I commented about needing only 14 to 16 functioning exposed mature leaves to carry a crop easily. At another point in the text I used the numbers 16 to 18 for considering distance@ 4" per internode. The discrepancy here is not error but rather depends upon and takes into consideration the intended lower leaf stripping practice.

I first started lower leaf stripping in 1976 and I spoke publicly about it, I believe, in 1977 at a WITS conference. When the audience did their mental time/motion/cost analysis and realized the cost per acre it seemed as if general apoplexy overcame everyone. "What kind of craziness is this that you are wasting our time with?" I was asked after leaving the podium. "Bullshit" was another term I overheard! Today, it is a standard feature in contracts.

But there were reasons for it. At that time we had two major problems facing us – the other problems hadn't really come up yet and were more "fine tuning" rather than major problems. Those problems were a) Botrytis and b) high PH along with high acid. Keep in mind that then we were still dealing with the Davis "haystack" vineyard and the new system was not operational.

The desiccated and senesced leaves in the heart of the vine created a high moisture area after rain or irrigation. They radically slowed down drying and they created barriers to sprays – dead leaves often "glued" themselves to clusters. Monterey was "Botrytis Central" – a constant threat and problem. Physical removal seemed an obvious solution though expensive and difficult (the same thing!) in the hay-stack system.

The pH business was a little more arcane. The basal leaves (the most mature) contain the highest concentration of po-tassium. Under the conservation of energy concept when the leaves senesce or begin to go dormant the vine transfers the good stuff out of the leaves into the canes. The important thought is the "senesce" part – its not just fall induced dor-mancy but senescence or decline from any cause. Shading out will cause senescence – consider all the dead leaves in the heart of a "hay stacked" vine in midsummer. Then also consider the seed hormonal system in those grapes not yet harvested mentioned earlier sitting nearby those desiccating leaves saying "send me the goodies". Boom – potassium goes to the fruit. What if – as a *remedial* act while we redesign – we removed those leaves while *green before* any decline? If they are physically gone they can't transfer the potassium and their loss won't hurt our CH production appreciably because they were going to decline anyway. At the time I called this the "Potassium Drive" because it drove up pH. A few years later Dr. Bolton at Davis did a project confirming the trans-fer, if my memory serves.

A minor third reason at that time was on Pinot Noir to attempt to increase anthocyanins in the skin to darken color from more light exposure.

These first two were the main reasons for the first work. The next step was to incorporate these recognitions into the factors of the new vineyard design that was in its germinal stages. That was done. The vertical positioning kept the bas-al leaves in the light thus avoiding shading senescence while fruit was still on the vine. This eliminated, or reduced, the improper early transference of potassium from the leaves to the fruit.

The vertical system also avoided the denseness of the "haystack" allowing rapid drying. There were no dead leaves holding moisture to the grapes. There was air movement both from ambient and sprayers.

Later, by 1982, we could point to the effects of light upon the flavours generated within the fruit. We used leaf stripping to both open up the canopy around the fruit area for even better spray penetration and coverage as well as light penetration for desirable flavour development. Further, this system used the natural precluders of disease – air and light – to reduce the incidence of diseases. When diseases came, or pre-emptive sprays were applied, no longer was large equipment pulled by massive high horsepower tractors needed to achieve penetration and coverage of the old style. Smaller units could be used thus reducing soil compaction and per unit of production capital costs not to mention fuel and chemicals.

I see many growers following the old rule of "if some is good, more is better". My personal belief is that this is not well advised in leaf stripping. We all have seen vines with their fruit hanging completely exposed. This has many drawbacks in my mind. Direct sunlight on the fruit can/will cause excessively high temperatures within the berry during hot spells and/or during the afternoon hours. The high temperatures will cause the malic acid component of TA (total acid) to drop precipitously by metabolizing.

The rate of Malic metabolization is a positive function of temperature. Witness the low Malic acid component of hot climate grapes. Preserving useful amounts of Malic is important and the reasons are discussed in the winemaking section. Those high temperatures can also yield a "cooked" flavour to the grapes changing the flavour complex substantially. A big loss is something I call "fruitiness" – a fresh lively component hard to define. Sun-burning of developing fruit will lead to "scar tissue" forming which has no flexibility and can split when the berries swell. These splits will leak juice providing a buffet upon which to feed for a wide range of bad things that can destroy the subsequent wine. It can also reduce tonnage but given the bad wine aspect that might not be such a bad thing.

There must be reasons why we grow grapes in cool climates otherwise why not grow them in hotter climes only such that higher yields could be achieved? By fully exposing the fruit to direct sunlight and heat throughout the day one shifts the nature of the fruit away from cool climate influences toward the hot climate nature. Heat is heat. Is that what one wishes? I think not!

In the fall as the temperatures of the Central Valley begin to subside the pressure differential between there and the coastal areas reduces and, thusly, the winds begin to diminish both in velocity and duration. In August a "hot spell" in Monterey can be 85 – 95 degrees while in October it can hit 105 degrees though not normally! The month of October and well into November is the common harvest period for Monterey. Anticipating this situation we like to leaf strip beginning *after* fruit set and pulling leaves only on the southeast side of the vines in the fruit area. The northwest side of the vines are NOT leaf stripped leaving them as protection against the afternoon sun and particularly against the October afternoon sun. During any hot spell we do not want that afternoon sun on the fruit. The mornings here are quite cool even in a hot spell and they supply us with all the light on fruit we need as well as all the light needed for fruit bud differentiation. We get sufficient anthocyanin concentration with this exposure regimen. I will point out that fully exposing most red varieties to direct sunlight will result in *reduced* anthocyanin concentration even to the point of "bleaching". The relationship between light and anthocyanin concentration varies by variety and understanding that is a function of grower/site experience. About 3 to 4 weeks before harvest we do strip some leaves on the west side of Cabernet to *increase* the heat.

Another factor in the lowering of the cordon wire was a thing called "moment arm". Basically this simply means that the longer a lever from the fulcrum the less pressure it takes to operate. Stood upright, if the "fulcrum" is the ground then the higher the trellis *weight* the easier it is for the trellis to go over. My proposal was going to create a "barn door" affect of wind resistance and a "lifting" of air by mechanical means (trap wires)! The wind pressures would be at the top area of the stake plus the weight of the fruit lower down would create an additive component. The higher up (or further away from the fulcrum) the fruiting portion was the less the pressure needed for failure. By lowering this unit down we enhanced the trellis strength from the haystack system. This "fulcrum" is actually the bend strength of the steel stake at or near ground level. Quiedan developed a rack that held a steel grape stake to which could be attached at various heights a wire that connected to a scale and then to a hand winch with which one could put tension until the grape stake bent. The side strength of the stake could be easily measured. Dan McNamara at Quiedan developed a metal grape stake that fit these strength needs. As an aside, I repeatedly observed him demonstrate the strengths of various competing stakes on the market. Many times I saw or have heard of farm managers buying stakes of 25% lower strength for 2% less money! I always wondered if the owners knew this! Strange economics but I guess it made a manager look good for awhile – until the stakes failed. No problem – by then the manager was gone or memory had dimmed. The owner suffered the loss. This is no small loss! The cost of trellis installation is a *huge* portion of the *initial* cost of developing a vineyard. It is even *more* costly when it involves removal of the old AND installation of the new! I have seen (over and over) managers save a few pennies by putting un-protected light gauge metal into clayish soils and have their trellis systems fail within ten years. I've seen one system like this fail in four years when the drip emitters were located *at the stake* ensuring that the stake was always wet! Of course, the stake supplier is always blamed – it is forgotten that light gauge with no protection was chosen against advice. Let me give you another warning. If an oil or gas pipeline runs through or near your land be aware that the oil companies put a slight charge on their pipes to *attract* electrons to forestall corrosion. That charge can "pull" from your stakes if unprotected! Non-operator owners – take note. What appears to be a decision of a few pennies is really one of mega-dollars! There is an old saying that nothing is more expensive than a cheap lawyer that also applies to your advisors on trellising materials.

The first formal trap wire holders were made by Quiedan with assistance from Terrell West. They were small crossarms bolted to the stake with lengths of 4", 6" or 8". There was a hole and groove at each end to slip the wire into. When installed the vine row looked like a small Appian Way of crosses waiting to crucify miscreant gophers or such. From my point of view there were some difficulties with them. They were not really easy to slip the wire into and out for one thing. Further, they were expensive. Not the cross arm itself really but the cost of installation. Power equipment, tractor and crew added up the cost. One day I remembered hanging Christmas tree bulbs on the tree with wire hangers, one end with a tight loop and the other with a hook. I asked Dan McNamara if he could form such a thing out of steel strong enough not to bend. He did so and the wire holder for the trap wire that simply attached to the fixed trellis wire was born. Fast, quick and inexpensive, it quickly replaced the little cross arm. There was a small problem with it early on. There was a tendency for it to come detached from the fixed wire when mechanically harvested and ending up in the grapes. That was fixed by tightening the loop with a pair of pliers when installed.

The timing of leaf stripping is, I believe, important. Not too many years after I began publicly speaking about it – maybe ten or so – others began to get into it – experimenting. I know of instances in warmer climes where the "experiments" were conducted in August and the fruit fully exposed. Of course, the fruit burned and the wine horrible. I heard several presentations on the subject and the "conclusion" was that it was not such a good practice. I commented after a couple of those that if I took my shirt off for the first time in the year in the month of August I'd probably burn too! Fruit that develops in light has a tougher skin – it acclimates. Still, some leaf protection is needed and the hotter the climate the more protection needed. Judgment is still required – it is a technique sort of like fire. Used right it will warm your house. Used wrong it will burn it down!

Conversely, I see many people leaf pulling before bloom and set, often when the shoot is out just a few leaves. That is certainly a time when it is easy to see and easy to get to. But "easy" isn't always what its knocked up to be. I have several problems with the procedure however. At this stage the shoot is growing on the *stored* energy noted earlier. To remove the leaves before much formation would, of course, not waste much energy. To remove them a bit later when they are nearly mature would waste more energy. All in all it would delay the formation of the canopy somewhat. It is in the nature of the vine to bloom early in the year and to draw upon the food being produced by the relatively few mature leaves in addition to the remaining stored food. We are not yet clever enough to measure the amount of stored energy nor can we measure the contribution of younger leaves nor do we know the new energy needs and form thereof the blooming clusters require, if any. Given this ignorance I prefer to leave everything up to Mother Nature until the fruit is set. We do know that zinc and boron deficiencies will seriously affect set and number of seeds within each berry. What else is needed? The grower's income depends upon that crop *and the next* and messing around with that many unknowns is folly.

The reason I italicized "and the next" above is that about the same time as bloom and set are underway the vine is deciding whether or not to make the basal (earliest) buds fruitful and how many cluster initials to tuck in there. This process is nutritional in nature and I think depends upon the total energy available at the time. That is, the combination of stored energy and new generated energy in *excess* of that required for shoot tip extension. Again, leaf stripping before set reduces this total particularly if the preceding year was a large crop year or if fall weather came early. The observed "big crop, small crop" syndrome was common in the past as it is today. It is a little more complex than that, though.

The spring and early summer weather in the cool climate zones can be quite cool and windy in some years while others can be less so. In those cool, windy years the early growth will be delayed somewhat. The new CH production is a function of the number of leaves that have developed *past* the demarcation line of energy users to energy contributors. In that case the available *total* energy available in the vine will be lower than if conditions where opposite. Further, in this case the vine operations are slower and the rate of transfer ability of energy from stored to available is slowed. There seems to be a mechanism that the vine uses to sense these fine differences in available total energy to differentiate between vegetative, partially fruitful and full fruitfulness. We have all seen this "none, one, two or three" cluster formation. If the old method of shutting down the farming (irrigation) when harvest is complete is followed then that 30 – 60 days of post harvest photosynthesis potential is lost and the absolute value of stored energy is reduced almost to depletion – it was used to ripen the crop and not replenished. Combine that with a less than desirable spring and – even if the rest of the year is optimal – the basal buds will not form the desired initials. Those that do form are small in size and berry number.

As discussed, fruit bud differentiation is a function of available energy AND temperature *at the time* of differentiation which is sequential out the new cane beginning with the buds first formed. In some areas of California's coastal regions the spring weather can be often, or regularly, cold and very windy, said wind blowing steadily through most of the day for days on end. Later, the winds diminish in duration and temperatures rise. During the period from bud break until new canes gain sufficient length to form a wind barrier and lift the wind over the vineyard the young leaves are severely buffeted by the cold winds.

During the early cool period the *rate* of photosynthesis is reduced from the temperature factor effects upon the chemical process. Further, the unimpeded wind strips the "boundary layer" of air adjacent to the stomati causing the precipitous discontinuity in the partial pressure thus resulting in closure of the stomati throughout much of the day closing off the gas exchange of photosynthesis. The combination of these two factors in these conditions can have the effect of dropping the available energy/temperature complex below the critical level for fruit initial formation to varying degrees – zero clusters, one cluster only, and/or very small sized cluster initials. This phenomenon is most common with early pushing varieties such as Chardonnay and Pinot Noir. Later pushes such as Riesling or Cabernet Sauvignon are seldom plagued in the same conditions though Riesling may display the syndrome by one cluster while buds further out will be twos or threes.

In these conditions one finds that the buds further out the cane – say beginning with the fourth or fifth bud – are replete with fruit initials, usually doubles of large size and some triples. If one is faced with this syndrome a change in mindset is required to viewing the spurs on the cordon as cane sources *only* and any fruit there from is simply frosting. A shortened cordon – not reaching the next vine – is admissible while maintaining at *least* eight spur sites though ten is better. This provides two canes each year while changing cane spur sources each year and allows spur site renewal and maintenance. The two retained canes are then brought up to the second wire and either tied out that wire over the gap areas *OR* are brought up and over the second wire and back down to the cordon wire to which they are tied. A little counting and arithmetic will show a much larger bud count than straight spur pruning, the buds will be fruitful and fruit separation will be reasonably accomplished in the vertical plane. Any failure of those exposed buds to become fruitful is irrelevant as those will be disposed of the following pruning season.

The phenomenon will be expressed in very subtle ways upon inspection. To explore it for one's vineyard study carefully the cluster counts from basal sources and those from buds further out the cane. In addition pay attention to size of each and presence or absence of millerandange (Hens & Chicks).

If one irrigates post-harvest then that period of accumulation of stored CH works to offset possible effects of the future and unknown spring weather. Many scenarios can be constructed based upon the "nutritional control" concept of fruit initial formation. Responses to observed condition can be formulated in advance by playing "What if" games and writing down the procedures to be followed if the given scenario occurs. The goal is to adjust to conditions and use farming techniques to offset any negative climate conditions. One is back to that equation making farming practices a variable.

Let me give an example of this process at work. Some years ago – perhaps ten or twelve – there was a spring that after about a foot or so of growth on Chardonnay the weather turned very cold for quite a long period. The Monterey vineyards turned yellow –golden yellow, bright yellow. One could drive down Highway 101 and admire the beautiful shimmering gold-foil appearance of vineyards on either side of the valley. This was not good given the above nutritional discussion. The color of the leaves is a function of the *density* of the chloroplasts – the little cells that do the photosynthesis work. Green is their color and green is good. Yellow is not good especially at that stage.

Grapevines have some difficulties with certain forms of fertilizers in cold conditions. The plant will pick up the nitrate form of N preferentially over the ammonia form but if the ammonia form is the available one it will pick it up. The problem here is that it can only use the nitrate form for growth so when it picks up the ammonia form it must convert it to the nitrate form. That is "work" and work takes energy! For awhile that conversion can really deplete the vine and debilitate it. A further problem is that in very cold conditions – soil conditions – it has severe difficulty picking up *either form*! Curiously, it has no problem picking up calcium molecules in cold conditions.

When the first signs of the "yellowing" occurred we applied Calcium Nitrate and irrigated it in. The vine will latch onto the calcium molecule dragging the nitrate in with it. In cold conditions one will see a response to calcium-nitrate within two days. During the episode noted above we were the only "Green" vineyard in the valley. Many growers toured our road looking. Why the differences? Most growers use UN32 almost exclusively when they fertilize. Back then, if memory serves, UN32 was about 40 cents a unit and Calcium Nitrate was about 1 dollar or a dollar ten per unit. We had used CN for spring application only for perhaps twenty years knowing we had cold spring soil normally but that year was the worst I had seen. Later in the year we will also use UN32, if necessary, when the soil is warmer depending upon soil bacteria to do the conversion for nitrate – not the vine.

As a further aside, a long time back Dr. Jim Cook was the nutritional guru at Davis and he did a lot of very good work but some of it is dated. The methods were good advice for the time but some are not good given our present understandings so take caution when reading his books and writings. One of his recommendations then was to apply a year's supply of N fertilizer in the winter using rains for incorporation into the soil. Today we have sprinklers and/or drip to use for incorporation. I am of the opinion that winter application is poor practice. First of all one has that cold soil problem with the vine. Many writings assert that "N is N" and many growers followed that dictum opting for the least expensive form. That is simply not true. There are different responses to different forms and different residual effects upon the soil such as acidification to mention one. Notice the "least expensive" terminology – they are all expensive. Winter application gives the grower poor control unless he knows more about predicting rainfall than I. The type of soil one has is also a factor as to timing of application. The vine is not really needing an N source until one has perhaps at least a foot of new growth. Too much N at bloom can cause poor set and shatter and excessive internode length – that is, too much vigor. The Ventana has sandy, gravelly soil so it responds relatively quickly compared to heavier clay soils. At Ventana the rough program followed was to use foliage feeds of zinc and Boron along with mildew sprays to achieve full final dosage just at commencement of bloom – usually two or three passes by that time. We also included a "kitchen-sink" nutrient mix in that last "at bloom" stage. Late in bloom and around set we would apply UN32 or Calcium Nitrate depending upon spring conditions. Later in the summer the last portion of the year's needs would go down. The sandy soil would allow N to be wasted if all were applied at once. Calcium Nitrate leaches easily while UN32 has more soil staying power and is the better mid summer tool, if needed. If any micronutrient deficiencies develop they would be addressed with foliage feeds along with the mildew sprays.

Calcium is a very necessary element for strength of cell walls and thus provides resistance to invasions. Nitrogen, to the contrary, in excessive amounts leads to weakened fruit cell walls and increases susceptibility to invasive organisms. We know that the leaves of apple trees are very receptive to calcium uptake therefore it is reasonable to assume that grape leaves are so inclined. It is my observation that this is correct. I have noted that in many areas of fame for quality in the world the vines grow in soil abundant with calcareous material. I utilize calcium in my foliage applications in conjunction with mildew sprays and general foliar applications until mid-season.

There are, of course, other formulations of fertilizers available. I have difficulties with each that I won't bother with here but I know many of them find favor with other growers. It is a complicated subject and good people can arrive at opposing opinions.

With all the above being said about controlling *excessive* vigor one does wish to get the canopy built as fast as possible. The sooner that canopy is in place and wired up the sooner the wind effects are reduced and the full "factory" is up and running. Personally I prefer not putting wires up while bloom is occurring unless the shoot growth is interfering with a needed spray. I have no particular theory to offer on this. It is just a gut feel that I would not like to mess around with the vine while it is busy setting fruit – something I sincerely wish it to do!

Once the shoots are of sufficient length to "wire up" and set is complete leaf removal can begin. There are sufficient mature leaves at this time to carry the load and the temperatures good enough to ensure good rates of photosynthesis. Even if there are coldish winds the temperature within the tunnels will be higher than ambient because they are out of the wind stream.

The molding part of the "Mold and Hold" concept should be concluded by late July or very early August at the latest. The mindset should shift to the "Hold" part. The farming techniques associated with this part are significantly different than when one is pushing for growth. Remember the old dictum – "Dormant pruning is vigor inducing. Summer pruning is dwarfing in nature." Leaf stripping is dwarfing in nature. Tipping, shearing and green cane cutting are all dwarfing acts. Water can be managed to avoid more growth – a little stress is good. About 30 – 40 days before expected harvest tip growth should be stopped or they should be cut. The "holding" part is critical to the development of the desirable flavours in the finished wine. Lignification of the cane at this time may also be involved in flavour development. They seem to go hand in hand.

If the year has been one of less than desirable conditions or if the pruning was just screwed up such that the shoot growth is simply not going to achieve these parameters green dropping should have occurred when the fact was first noted earlier. However, green dropping can be tricky. If the dropping is because of stunted growth – over cropping – then the earlier the better. The vine needs time to correct itself. Any delay in this instance will result not only in improper shoot growth for *ripening* (not dehydrating down to sugar) the fruit but will also lead to failure of buds to differentiate for next year. The basal buds are quite probably already affected and that should be considered for the next winter's pruning. Later, but before pruning, bud analysis should be done in order to evaluate the situation before cutting.

However, if a *slight* overcrop situation is faced that is not necessarily a bad thing as it will help control over vigor. This is a judgment call. In this case green dropping should not take place until veraison is complete or nearly so. Also, if a grower is faced with a winemaker who, because of beliefs or mythology, wishes crop size reduced below the proper balance – and he is properly compensated in some fashion – then such green dropping should not occur until veraison is nearly complete. The reason for this is that if excess energy is available shoot growth may be difficult to stop. Also, the vine will try to compensate by swelling up the berry size if done before veraison. The object in fine winegrowing is to keep the relative berry size small thus increasing the skin to juice ratio for extracting essence – particularly so in reds. Big berry tonnage does not lead to the best wine. Notice that the *composition* of "tonnage" is critical to wine – not the absolute tonnage itself. Small tonnage with plum-sized berries equals lousy wine.

Those gurus who assert the mantra "smaller the crop the better the wine" actually worship a false god not to mention displaying a certain sad level of ignorance. Strong statement? Well, perhaps, but let's explore a bit. Please recall above I pointed out the equation wherein climate was, historically, the only variable. With one variable – climate – one would *rationally expect* that there be good years and bad and in-between according to climate differences. And, in reality, that is exactly what we have seen in history. If we look at an area with long history of production and of major interest by people we can look for general concordance. Bordeaux, happily, is such an area and many self-professed gurus and wine scribes can easily come up with a consensus list of superb vintages of the last century. Such lists are even in print. There is, nicely, also a book first printed in the late 1800's periodically updated to the present, that covers a variety of details on vineyards and districts including – yields! The book is titled "Bordeaux and Its Wines" by Feret et Fils. It usually is printed only in French but, from time to time, in English. One might find it of interest to compare the list of superb years for wines with the lists of yields. Quelle Surprise! It is often so aggravating when facts meet beliefs! Joseph Stalin is supposed to have observed "the irritating thing about facts is that they won't go away"!

Another example story may be instructive. Perhaps twenty years ago or so Professor Helmut Becker of Geisenheim fame visited here and lectured to growers and winemakers. This was a time when Riesling yields in California were around 4 tons per acre. Americans were struggling to make quality Riesling wines – flavorful not just sweet with Muscat added. The yield mantra was in full force and I was fighting winemakers on the subject (Ventana Riesling wines were consistent gold medal receivers – now 30 consecutive years). During the question portion I asked Professor Becker a question that I thought would be a revelation to our crowd. I asked "Dr. Becker, isn't it true that Schloss Johannisberg, over the last decade, has average over 8 tons to the acre and Shloss Vollrads over 11?" He looked at me for some time, then with a smile nodded his head 'yes', looked away and asked for the next question. During his talk he had spoken of low cropping. I had selected the two most famous quality estates in the most prestigious German region both of which looked downhill over his university! It seemed not one of my compatriots got the importance of that question – and of Professor Becker's response! I had hoped to attract serious attention to the different viticulture paradigm and vine densities that had so much importance for wine quality and economics – to no avail. Understanding and acceptance was still far in the future.

When tillage begins, other arts follow. The farmers, therefore, are the founders of human civilization.
—Daniel Webster

Section 3L

HARVESTING

With the old system in our windy, cool climates there were times when primary bloom took as long as sixty days though for most fruit it was more common to last the thirty to forty day span. That, also, was entirely too long giving fruit far in excess of the desired fruit ripening range. As mentioned elsewhere we used selective hand harvesting as a makeshift solution while re-designing the entire system to reduce that deviation around the mean. Now, with the new system, we fall normally within eight to twelve days though extra coolness can stretch it two or three days. The mentioned time span is within a block according to pruning time not necessarily according to calendar – that is, elapsed time not an absolute date. The "veggie" component on white vines has disappeared and is no longer even considered a factor in thought. The young growers and managers are shocked when it is mentioned. They look at you confused and wonder what the old geezers are talking about! Unknown to them, it was a major element then that cost the Monterey region many, many millions in dollars and even more in image in the marketplace. Once solved in whites it took perhaps fifteen years to overcome and to die out in people's minds.

There was a love affair with the mechanical harvester then to the extent that many corporate growers did not even possess gondolas, trailers and other equipment to allow them to hand harvest. Cesar Chavez was organizing farm workers and growers were scared to death of him. The idea of farmworkers striking during harvest with the grower having no alternatives gave nightmares. Even without the fears there was question about the availability of sufficient labor to harvest the crop from the huge expanse of winegrapes on time. The mechanical harvester was the answer.

The early harvesters were cumbersome machines with thick carriage units of "elephant ears" below the vine to catch the falling grapes. This required "high" growing of vines on a trellis. Free-standing vines were/are not amenable to machines. In the late sixties the normal vineyard design in Napa and Sonoma was the free-standing vine each with its own stake. In 1973 I went to Napa to look more into what this "grape-growing" stuff was all about. The big discussions going on among growers was how to convert the vineyards to the trellis system without losing a crop year completely.

For the first twenty or more years of its life the mechanical harvester depended upon straight fiberglass rods beating upon the vines. The tips of those rods approached extremely high velocities and, in my opinion, were severely damaging to the vine and to yields. I succumbed to the paranoia and purchased one of the machines mainly intending it as an "insurance policy" in the event of labor difficulties which never materialized for me. It did provide me the opportunity to study the actual operating defects of the approach. It was also used quite heavily on other people's vineyards as machine harvesting was their accepted mode of operating. It was used minimally on the Ventana for reasons which I will clarify.

The operating theory behind the machines was to transfer vibrating energy to the cluster via the rods from the harvester. Theoretically it was not intended to "beat" the grapes off by smacking them. To properly transfer the energy the trellis system *AND THE ATTACHMENT OF THE VINE THERETO* had to be solid. If there was flexibility in that attachment then that would allow absorption of energy and not transfer it to the grape. The effectiveness of the harvester was directly affected by the diligence of securing the plant to the trellis. It still is today with the new head.

From the mid-seventies through the mid-eighties I was buying and gathering tiny amounts of fruit from various vineyards throughout the valley to map the flavour differentials – and those differentials were great. In the course of this activity there were some growers who offered to give me some fruit if I would make wine and tell them about their grapes from which it was made. In other instances I would buy grapes in quantity and sell the grapes to other wineries. We were equipped for hand harvest. I would make an offer of a small amount per ton, we would "glean" the ends of rows and, in many instances, the field and resell at much higher price. The "field" instances were the most interesting because we observed that where the canes were not secured tightly the cane would flap around when the harvester went by and no fruit came off. The "custom harvester" companies paid no attention and corporate managers paid little attention. When a complaint was made the response was to increase the beater speed thus damaging the vine even more. The harvesting problems weren't so much the machines as they were of failure to understand structural matters.

One area of the structural aspect was allowing shoot growth in the yoke area and deliberately keeping spurs in this area – as is still common to see today. Those shoots and fruit in that area created a dense "basket" arrangement that not only harbored disease in its denseness but also made it difficult for the harvesters to shake the fruit out. Growers, following the machines, would see that fruit still on the vine at each stake and would instruct the driver to increase beater speed until they saw no more fruit. Again, the result was more vine damage. By simply making sure there was no growth in that area much of the vine spur and trellis damage could have been avoided.

Besides the obvious vine damage of broken out spur sites, broken cordon arms, trellis damage and cracked canes there was a far more insidious effect of machine harvest not so apparent. Over the years I repeatedly demonstrated this effect to no avail – primarily because of the lack of an alternative for large farm management companies. The demonstration involved hand picking two rows and machine picking two rows adjacent. Then I would turn on the sprinklers. Within a week the hand picked rows would still be lush and green photosynthizing away while the machine picked rows would be defoliated except for a few young leaves of the tips of the canes. The reason for this was that the violence of the rods had caused the vascular tissues of the leaf petiole to break! After the machine had passed the vines *LOOKED* green but, in fact, the leaves were no longer functional. The same effect could be observed following a rainfall post harvest. Up to sixty days of photosynthesis was lost. In each instance that we tracked the yield from the rows in the subsequent year the handpicked rows out-performed the machined rows by a minimum of over one ton per acre – usually closer to one and a half!

A different sort of harvester was also developed about this time – a trunk shaker. While I have had no experience with it, it did not break out spurs, yet – it really did not gain wide acceptance.

Somewhere in the eighties a grower from Napa acquired an existing vineyard in Monterey. This fellow had been in winegrapes his entire life – both here and growing up in another country. He called and wished to hire me to consult - a practice which I tried then to avoid. Most of my thoughts about viticulture then were generally considered heretical and blasphemous and I preferred not to engage in that sort of work. I already knew I was crazy and didn't need to be regularly re-informed of that fact. I was too busy sorting all the stuff out and didn't wish to waste my time. The fellow persisted and so I gave him one-half day, told him I would speak with him and named an exorbitant fee hoping to dissuade him. We met in a coffee shop in Salinas. In the course of briefing him on cold climate and windy viticulture he and his associate were not terribly well receptive to the various points I was making and the idea that he should consider a plan of re-developing his 12 foot spaced vineyard was rejected out-of-hand. Then I made a comment in passing of no great importance to me but one that fired him off. He made some comment about wasted money – my fee. My comment was so contrary to the conventional wisdom that he thought I was nuts. The comment was "If the price of grapes is $1,000 per ton or more it is cheaper by far to hand pick". After his response I simply looked at him for awhile and then asked him to bear with me for a moment and to take a fresh page of his yellow pad. I asked him to draw a vertical line down the middle which he did. I asked him what was the going rate for custom harvest and he responded $100 per acre up to 4 tons per acre and $25 per ton over. That is $25 per ton cost. I then asked him if $110 per ton was a reasonable cost for hand picking (which was high then) and he concurred pointing out the huge difference. I said "right but we're not done yet". I asked him when he hand picked if the winery price included the stems. He looked at me strangely and said yes. I asked if stems were still on the vine when machined – yes. I then asked what percentage of a ton of handpicked fruit was stems. He didn't know. I said it varies by variety or by year (size of berries) within a variety but on Chardonnay it was about 5%. He said he'd been in this his whole life and had never heard of such a thing and how did I know? Simple, I said – I'd bring in a ton of grapes, destem and weigh them. So, at $1,000 per ton 5% is $50 per ton of stems that I get back for hand picking which leaves us at $60 per ton hand versus $25 for machine. He said still machine is cheaper. I said "now – a quirk in our accounting system and thus our thinking. Next spring how much an acre is spent in trellis repair if you hand picked?" He said usually zero. I said "right!" "Now how much if you machined?" Depending upon the operator and vineyard installation it varies but we agreed that $200 per acre was a fair number and more if wooden stakes are present. That added $50 to the machine tonnage price and put the numbers at $75 machine and $60 for hand. Then I informed him of the 1 ton per acre effect from reduced carbohydrate storage, said "add another $250 (one fourth of $1,000 per ton lost income) per ton to the machined ton". Even without the esoteric lost potential yield the hand was better than machine. This analysis did not even bring to bear the lost future yield due to spur site loss. Then I left.

In those days growers were concerned about losses from juicing by machine. A favorite touting by the academics was that machines could increase yield by not missing clusters – Gewurz was the given example as their small dusters "hid". Notice that the new system eliminated this and all the "hunting" through haystacks.

Today the same sort of analysis should be done by each grower however a whole different set of parameters are extant. The new picking heads do not do any where near as much damage, labor is far more "iffy" in quantity available, the heavier yields per acre mediate the higher labor costs, etc., etc. There is a caution in "time" though. With winemakers pushing for higher sugars the window for harvest is much more narrow. Handpicking will probably become the norm only for the highest grades of wine. Still, I prefer it on reds if for nothing else for its cleanliness – that is, the lack of leaves and the ability to skip any deteriorated clusters. Hand picking is not as expensive as one may think at first blink!

The new system was designed to make mechanical harvesting easier and less damaging to the vine. The "beater" cycles could be much, much lower thus reducing significantly the breakage of spurs, no jam-ups in the yoke or bend area to induce higher beater rates and no mass of canes to beat through. At the time I was regularly asserting that if the machine boys didn't come up with a new picking head when it was needed for this system I'd have to design one myself. In 1988 or 1990, in Germany, I was taken into a "restricted" warehouse to be shown a new innovation in picking head design by BRAUD. As it was being shown to me they started explaining it but I stopped them saying "I understand it – it's brilliant, and such a simple solution". It was the bowed-rod concept. It went on to receive the Europe gold medal for equipment innovation. No more super-sonic whacking by rod tips - just the number of bowed rods that were needed to fit the vertical design of fruiting area. Another problem solved – not by me but by someone.

As time has passed the collector units have thinned down enabling harvest much lower down toward the ground. That aspect is no longer a real issue thus trellis and cordon heights should be a function of the other points. Properly adjusted and operated the new machines do an excellent job of de-livering clean fruit absent trash and leaves. They also do an excellent job of sorting out botrytised or rotted fruit – if dry. Even the old machine did this at times. In 1978, I believe, another vineyard had a major attack of botrytis and decided the field was a walk-away. I knew a winemaker who wished botrytised fruit. He and I were in the field when I put a ma-chine on a row. After a ways down the row he looked in the gondola and yelled "Hey, where's the botrytis?" I looked and there were only the sound berries there! The machine fans had pulled the botrysized berries out. We were picking the wrong fruit for him! We switched to hand.

The new system was designed to address the deficiencies in the harvesting methods of the time. In the case of hand-picking it presented the fruit directly in front of the worker at ergonomically correct heights and eliminated the "hunting" through foliage and canes both of which drastically reduced costs. The "hunting" aspect was not only time consuming but was also fatiguing to the worker as the day progressed. In the machine case reduced RPMs reduced damage to the vine, reduced canes jamming up the machine, reduced damage to the machine and associated costs, reduced leaves jamming fans thus costs – and improved quality, etc. etc.

With the now-called VSP system (and some of the various others) only the number of rods to work the fruiting area are needed so that no rods impact on the foliage-only areas. The very low vibration rate needed reduces or eliminates the vascular damage to petioles thus allowing subsequent CH production and storage. With these aspects under control the very beneficial aspects of machine harvesting can be utilized for first rate white wines – night and speed. Night picking allows cool or cold harvesting. Grapes came off easier if cold than warm thus lower RPMs are required and reduce cooling needs in the winery. Speed is important to reduce skin dwell-time. At Ventana we could at times accomplish first grapes off vines to last grapes in the press within 2 hours. Of course, that is a winery-on-site situation.

*Every great advance in natural knowledge has
involved the absolute rejection of authority,
the cherishing of the keenest skepticism,
the annihilation of the spirit of blind faith.*
—T. H. Huxley

Section 3M
TODAY

As anyone can see by driving the length of California to-day a massive project of conversion to the elements of this design is well underway. For the first ten to fifteen years or so of its existence it was mainly a target of humor and was considered a curiosity to be only looked at and deprecated. Toward the end of that span – in the late eighties – things began to change. People began to visit the vineyard at night armed with flashlights and tape measures. Others came on Sundays. We did not interfere though we were aware of their presence. It was a bit humorous to watch the skulking about because for years I had been very open about the merits of the design and its affects upon yields and quality. These folks were locals and perhaps a little embarrassed about some of the earlier opinions expressed. Whatever the reason, if that is how the important information for the new viticulture had to be spread – so be it. There were also those who were very upfront, asked me a few questions and proceeded to install small vineyard plots.

A vineyard cannot be hidden so the blocks were there for all to see. Our equipment area was apparent. The equipment formats had drawn some attention post 1990. It was not until *after* that year that we solved that particular complex. Attendance at the enology and vineyard equipment show in Stuttgart, Germany solidified my thinking on the equipment aspect. I then utilized European tractors which were not satisfactory for various reasons – parts, price and timeliness of note. Not until John Deere's entry into the market in the mid-nineties did that area come to an acceptable finish. Construction of the other equipment was initially done for me by shops in the Central Valley or locally. Other growers were observing our on-going conversion of the entire Ventana Vineyard over those years. The counting of clusters and applying some basic arithmetic was raising some eyebrows. Humor was changing to thoughtfulness.

Thoughtfulness, though, could not carry the observers to complete acceptance. The design as they observed it on Ventana was just too radical of a departure from the conventional wisdom for it to be digested in one step. In fact, I stopped speaking entirely on the theoretical and mathematical aspects as too confusing of the issues and stuck to the mechanical "how to" commentary. It was easier for vineyard managers to see the benefits but the tightness of the spacing concerned them. Owners were an entirely different matter.

Owners face a different array of barriers beyond their initial own ability to understand and accept the concepts. Particularly when it was so divergent from conventional formula. Their explorations and observations of The Ventana strongly inclined them to the economics of yield of closer rows (more intensive use of land) but they couldn't stomach the full deviation from the norm without some small testing on their own lands. To follow the full leap of logic would also cost an investment in a full set of equipment in addition to the large up-front cost of installation. These first "copy" vineyards were only possible to successful established owners because of the banker element. There is nothing that scares a banker more than being in a lender position on an unconventional approach to agriculture. In fact, since 1985, agriculture itself was a victim of the American banking industry when it changed from the historical asset-based format to short-term cash flow basis. That was devastating to the historically asset-rich, cash-poor industry. At that time to bring forward an expensive revolutionary methodology within that maligned industry was to invite severe problems. I know – it was my lot for thirty years to teach the financiers! For a non-self financed operator to attempt an installation would have been folly. In my case I believe the bankers viewed it as a charity – aiding a crazy person.

And so it is that within a thirty minute circle tour near Ventana one can still see the sequence of evolution. There were two "firsts" – one at 9 feet between the rows and the other at 10 feet. Neither could at that time buy in to the Ventana cordon wire height so we see a hybrid about half-way between the California high wire and Ventana's wire height. A couple of years later the 10 foot owner installed another block at 9 feet. A couple of years later another grower installed at 8 feet and many others shortly followed. The "critical mass" for wide spread application was probably when Kendall-Jackson, on large acreage adjacent to and above The Ventana, tore out huge acreage of recently grafted over old-style vineyard and replanted to the eight foot number while it caught its breath and acquired some experience. That pattern spread north and south in conjunction with the afore-mentioned Phylloxera fiasco. Players in Monterey took the design to other lands they owned in other regions.

Meanwhile, at the later stages, there were others who were faced with the existing 12 feet rows as I had been who could not tear out and replant. They, too, began to interplant between the rows at 6 feet spacing.

For perhaps more than ten years now one cannot borrow money from long-term lenders (read insurance companies) to install an old style vineyard. The elements of the new system must be incorporated. In a slightly modified form for the conditions it has even made its way to the Central Valley in *very* large newer plantings. Different conditions can dictate those modifications but to make them properly one must know what and why. The system has made it to Europe as well as Australia and New Zealand, Argentina, Uruguay, Chili and probably other areas unknown to me.

Bankers and investors had had time to chew their cud and ruminate on the new system's costs – both installation and annual operating costs. You cannot imagine the hassles of all those early years with my crop financers. The *per acre* operating *costs* were so out of whack with conventional cost structures with which the bankers were familiar (and the comparisons with their other customers within their lending area) that it was a continuous bone of contention. Of course, cost was their only focus. I would repeatedly show the relationship between sales income, direct costs and *cost per unit* of *production* (which was substantially below their other borrowers) and they would assert that they understood. And then they would again raise questions about the cost per acre! Gotta love 'em. Teaching and conditioning them on new subjects takes time. It is the same with owners and investors. It takes time to adjust to an entirely new set of economics and habitual mindsets.

Far from the terrible labor pains of its birth, Monterey today is in the process of fulfilling early predictions of its greatness as a world recognized wine region. What it took was the development of a new paradigm of viticulture to allow its uniqueness to show and glow on the world stage. That was the clearly and often stated purpose of the work from the very beginning. There has never been a deviation from that stated goal of some thirty years ago. It required a complete re-thinking of the entire system and a willingness to cast off the shackles of the past lignified thinking and absorb the abuse offered by worshippers of status quo.

Some have held my feet to the fire for being so abrasive and, at times, demeaning in my assaults upon the conventional. I concur that I often was those things in my approach, perhaps even more than that. However, that was not by chance. In Monterey in those years we were observing bankruptcies and tear-outs – basically a destruction of our young region before it could reach puberty. As the answers became apparent to me it also was apparent that we had very little relative time to cause the changes that would save our region. I did try the behind the scenes approach but they either could not or would not understand the points of objection. To smile benignly and gently allow the ideas to slowly permeate and wait for the "old guard" committed to the past to simply die out was not an option. Ok, sure, Ventana could have dazzled but how the hell does that build a region? We had already seen articles damning Monterey "except for Chalone and Ventana". That was not a beneficial route.

Embarrassment within the academic culture is a powerful force and that force, among others, was the necessary tool to force change. It still is. To hold up absolutely ludicrous thinking and writing from the academic community to public scrutiny was warfare not normal to their experience. Many considered my attacks unfair and probably still do today. Perhaps this book will fall into that area also. I'm sure there will be defensive reactions. Goes with the territory. I was not concerned with the sensibilities of a few coddled prima donnas who were spewing misinformation that was damaging in nature. Their thinking was simply "in the way", an impediment to change that must be neutralized and the misinformation clearly shown to be that. Time I did not have. One cannot engage in that sort of mission by simple quiet assertion. One not only must be more "academic" than the academics but must take the battle into loud, open public turf where the subjects cannot be quietly buried. In other words, I was going to aggressively hold the old ways and their priests accountable. I did. After more than fifteen years of struggle (still ongoing in the theoretical level) we finally observed structural academic acceptance after forced by the industry conversions to the level of face – saving claims of parenthood. We practitioners welcome academia aboard. We do need some future assistance in elsewhere-noted areas.

In 1905, Albert Einstein had published a flurry of essays in an eminent journal that devised a revolutionary quantum theory of light, helped prove the existence of atoms, explained Brownian motion, upended the concept of space and time and produced what would become science's best known equation ($E=MC2$). The third-class patent examiner hoped that this work would lift him from obscurity and provide academic recognition. His sister later commented "But he was bitterly disappointed. Icy silence followed the publication". Such is academia, its guarding of personal turf and its resistance to change. Such it has always been.

In the early sixties I first read a tome called "The Structure of Scientific Revolution" (Kuhn; Columbia University Press) my copy of which became worn out later and needed replacement in the nineties when re-printed. The theme of resistance to change by academia is a repetitive one. When the growing mountain of evidence indicates the pressing need to go "public" (and the obvious fact that one can't hide a vineyard – it sits there for all to see) I knew the travails that lay ahead. Enjoyable? No! Necessary for the development of Monterey and California into greatness? Yes! So be it – those were the cards I was dealt and play them I must. There were times when I simply retreated into the bushes, licked my wounds and refused to speak out. That seemed, strangely enough, to anger my critics more than my earlier open, heretical and blasphemous speaking. Perhaps they just smelled tiredness and weakness. Predators do that. Eventually, though, the sense of duty would prevail and I would "gird my loins", pick up the lance and go again in the search of windmills to slay.

Happily for California the new system has been of benefit to those far removed from Monterey. The result has been a remarkable improvement in the quality of grapes for the making of modern flavorful wines wherever it is utilized.

The practitioners are still in the process of learning how to manipulate the factors to achieve the desired wine – if they are ever told what that is. Monterey faces challenges in the future because of a sever shortage of small wine estates each chasing their own vision of art and excellence. Progress and reputation lies there – in those works of art. As an indicator of Monterey's rise to excellence of quality is the presence of large industrial wine operations. Those types of operations do not come around when a region is in difficulty. They began showing up in Monterey since fifteen years ago when the solutions to difficulties were becoming known. That very indicator will also be one of Monterey's problems in its rise to prominence. To a very large degree the large corporate industrial wine producers and the large farm management operations catering to those interests are detrimental to that climb to the esoteric levels. It is not that they are not nice people! It is because the very nature of their business and its methods is not amenable to the expression of those sorts of wines upon which the highest regional reputations are built. It's not that they don't have the assets to work toward it if some corporate will to do so should erupt. Its not that at all. Look at Gallo's decades long efforts in that direction in Sonoma. No – it's not that at all. I think that those levels are achieved by restless spirit, fearless artists who are forever driven. It strikes me that that sort of person is constitutionally incompatible with corporate mentality and structure. Conversely, a good corporate worker would be driven bonkers if forced to work in conjunction with such a person. Yet – that loose cannon is where success lies. And failure. And the corporate world does not have an affection for the failures while that artiste is learning how to fly. Another structural flaw, as mentioned elsewhere, is in the corporate world's inability to give absolute control over growing and winemaking – that is, winegrowing.

Winemaking throughout the world has improved remarkably over the last twenty years and in particular over the last ten – awesomely at the industrial level. Artisan wines did not have as far to go but there, too, remarkable improvement has been made. Not so very long ago many of the wines made by the large producers of the south of France were such that one had to be born to them in order to get them past one's lips. Today, most are quite commercially acceptable. Americans buy many of these fancied-up in 750 ml bottles with pretty labels – but they are jug wines none-the-less. With modern techniques they are sound and tasty in many instances. I recommend Tuesdays – Tuesdays are often good vintages. The same thing can be said about our homegrown mega-producers.

Boosting the surge in quality of the world's wine and California's in particular has been a surge in the general quality of the grape. For the upper level of quality wines the new vineyard design has generated quality grapes far superior to the old California sprawl system's fruit. While in warmer climes some may argue their case (and lose I think) but in cooler climes there is more than sufficient proof of the efficacy of the new system in supplying spectacular fruit for outstanding wines.

Today, Monterey and California are producing world class and world competitive wines. Period! The New Viticulture has significantly enabled that achievement. Monterey is fulfilling its destiny. My work is done.

*Beware of taking any one thing out of its
connections, for that way folly lies.*
— Ralph Waldo Emerson

Section 4
THE MAKING OF WINE

In 1974 I went to Napa to see what this winemaking "stuff"
was all about as well as to observe their growing practices.
Napa was, after all, the Mecca to everyone in the fine
American wine industry. There was one dive motel just south
of St. Helena and a "hotel" in Calistoga. For fine dining one
drove to San Francisco. It was truly country – not the yuppie
upscale area it is today.

I decided that that fall I would make a few carboys of
wine. I purchased the few books that were easily available
and a couple from U.C. Davis. I did not study them deeply as
I was enmeshed in installing vineyard and tending apples in
Washington. Some home winemaker friends helped me with
some sulphite tablets, a handful of yeast and some general
advice. One in particular was very helpful – Ron Lamb of
Morgan Hill. I bought a hand operated crusher, some plastic
garbage barrels and – with my navy mesh nylon laundry
bag – set to work. (SEE PLATES 70, 71, and 72).

Peter Mirassou let me glean some grapes after his harvest
which we put into the back of a pickup after we had lined
the bed with plastic. We put the laundry bag into a garbage
can and crushed the grapes into it lot by lot. When full we
twisted the laundry bag like one twists a towel to wring water
out. It was tedious work but we did get juice. After settling
the juice we placed it in carboys, added yeast and secured
them with fermentation locks supplied by Ron Lamb. In one
garbage can we fermented red grapes then "pressed" with

the same laundry bag into carboys. I set the carboys in front of my desk in my office. Oh, what a wonderful smell!

Shortly thereafter Andre and Dorothy Tchelistcheff came to my office brought by the principals in my company. It was the first time I met Andre and Dorothy. After the meeting, on their way out, Dorothy whispered to me that I should move those carboys out of the direct sunlight as sunlight was bad for wine. That was news to me and I did so promptly. Those carboys turned out to be very interesting to me. The wine from the white grapes, when settled, was delicious and we sampled out of the carboys. Soon the wine was *not* delicious and shortly became ugly. Of course, I knew nothing about the effects of air! The red wine was another matter – it smelled of burnt rubber. It was not at all pleasant. A winemaker acquaintance came by and I asked him to smell or taste it and tell me what was wrong. He did so and told me "That's Pinot Noir and that's how it smells young". God, it was horrible. It was very apparent to me that I needed more time in the books and needed to ask a lot more questions. Not too long later Andre and Dorothy were back and at the end of that meeting Andre said something to me that sounded like "malo-lactic fermentation" or some other Russian phrase. At the next meeting with Andre I said "Andre – you kind of sandbagged me". He said "what?". I said "I looked up that malo-lactic stuff. There are only three American writings and they all have some Russian guy as the author!" He laughed. But I guess I passed his test and he decided I wanted to learn. Over time he would make little comments to me on a wine – each of which would lead me on to more study and new frontiers. Andre was so subtle and diplomatic in his way of putting things that one could miss his points if one didn't listen carefully and then dwell upon the implications thereof.

By the 1975 harvest I was ready with a few barrels, carboys and various devices but still with the laundry bag, hand crusher, garbage cans – *and* a cylinder of CO_2. Everything went well, the wines were good, went dry, settled and racked with CO2. The bungs were in place. The barrels were in a barn. A couple of months after that I had some friends with me and I wanted to "show off" these lovely wines. Lo and behold, the liquid level was about eight inches or so down, creepy crawly things were there and the wine was, again, horrible. It turned out that my troops – who had worked hard processing it with me – had been siphoning off small amounts. They also did not understand air. Well, we poured out the wines and the barrels became fireplace wood. Back to the drawing board.

By the 1976 harvest many more books had been read (including one from England on how to make wine from Dandelion greens and Elderberries!) and more barrels acquired. I also had some discussions with my guys about not tapping the barrels. For the '76 harvest I had acquired several different strains of yeast. The American academics were asserting that there was no difference between the yeast strains – a position they would persist in until 1985. At a lecture for home winemakers and small wineries at Davis that year (1985) the shift occurred and it was asserted that while there *may* be differences in young wine they would disappear with a little time. Now how would they know that if they hadn't done side by side controlled experiments? And if they had done those experiments then they wouldn't have asserted no differences right along. Along with that there is a curious line in "Table Wines" (1970) about flavor-producing yeasts generating less alcohol and another that asserts that American yeasts were selected for their alcohol generating capabilities! Most curious, these "academics". There is a not-so-nice story about this but I'll skip it.

The Harvest year 1976 is when we began the multiple yeast trials by settling a small tank of juice, racking it to a CO2 filled small tank for uniformity and then to a series of carboys each of which was inoculated with a different yeast. The racking step was done to assure that the juice was uniform – that is, to avoid any "layering" in the primary vessel which would have made the results questionable. The differences were dramatic! Rates of fermentation were different as were foaming tendencies. During the fermentation at given stages the generated aromas were different. When dry the aromas – in most cases – were different however in some cases I couldn't detect differences nor could my people. Fermenting to dryness was a problem for a couple of them.

Along the way I had purchased various lab instruments – pH meters, various hydrometers, titration apparatus for total acidity, etc., and either taught myself their usage or learned from others. I put in a "lab bench" in the old barn on Ventana (still there today) and that became my workshop for the next thirty years. It was in 1976 that I *started* to become cognizant of problems in the fruit concerning Total Acid (TA), pH (propensity of Hydrogen), Brix (degrees of sugar) and "ripeness". The numbers that I was getting did not jive with either textbook American numbers or conversion of European numbers. By "conversion" I mean because different systems are used in different countries and, for comparison, one had to convert to a common scale. For example, the French measure TA as sulphuric while Americans base it on tartaric. In addition to measuring parameters on Monterey fruit and wine I had started buying wines, both American and European, and putting them through the lab bench. By using fruit from older vineyards (Ventana was not yet producing a commercial crop) hopefully these numbers were not a function of juvenile vines and should be reasonably comparable to commercial wines from elsewhere. They were not. One glaring dissonance was the combination of high pH and high TA. Another was the across-the-board higher pHs of American fine wines compared to European – in most instances. The deviations were interesting.

For a long period I was of the opinion that Burgundy made the best wines in the world. Sadly it also made a goodly share of the worst! Unfortunately they had a nasty habit of "barrel bottling" so the products – though legally labeled – were not uniform. One never knew from bottle to bottle what one was going to get! Buying both young wines and older, it became very apparent quickly that the unpleasant Burgundy wines had high pHs – often up around 4.0, sometimes higher. The pleasant wines and sound older wines had pHs of 3.4 to 3.6, sometimes a touch higher. One could almost measure pH and predict whether the wine was any good simply from the number. Of the ones I measured it was absolute that the older high pH wines would not age nicely – they fell apart and deteriorated rapidly. The recognition of this relationship was very important to my thinking. If one combines this observation with the other mentioned problem (high TA) then one can see why, at that time, it was necessary to re-evaluate the vineyard system. For those who are not aware, the solution to a high pH problem in a wine is the addition of acid – normally Tartaric which should lower pH. If you already have excessive acid the winemaker has no room to work. Remember, this was a time before fancy tools like ion exchange columns or such. The grapes of the warm and hot climates did not have the high acid problem as the climatic heat metabolized off the malic acid portion and the tartaric dropped nicely. In fact, low acid was a problem there (and thus, high pH). I have been told that long ago a very normal practice was that when shipping a railcar tanker of wine to the east 5 gallons of Sulphuric acid was ritually added to each tank car. The travel would mix it.

One problem, obviously, was how to grow the grapes in cold climate such that the TA and pH were lower – lower into desirable ranges. As one can see, the idea of getting away from basal leaf senescence is contributing to a reduction in potassium transference. High Potassium is a major factor in high pH. TA is a different matter – a bit more complicated though part of the same conduct. Fruit that is grown in the deep shade is cooler than fruit exposed to light. Some light exposure can raise the berry temperature aiding acid reduction. However, that was not the basic problem.

Rigid beliefs in myths was a greater problem. There was a hard belief by wineries that later season irrigation was a "negative" and thus *forbidden*. There were contracts that forbid irrigation after August 1 and, if caught doing so, the grower would lose his contract. At the same time, remember, the only harvest requirement was degrees Brix. These two things combined resulted in dehydration of berries *down* to sugar concentration – NOT accumulating sugar up. Keep in mind that sugar and TA are measured as percentages of a *SOLUTION*. If one irrigates the plant absorbs water, the berries take in water and the sugar and TA – as a percentage – drops back in normal ranges for the level of ripening. Interestingly, the pH doesn't move a great deal after a few days.

There was a year in there (1980?) that the head of a large corporate grower operation north of me had sold me grapes for my comparison work and bulk resale. He called me and pushed me to pick asserting others had done so and were running 23.5 degrees sugar. I thought – no way could that be right. I went, I looked, I called and said irrigate! Now. The TA was 1.4! He resisted – he thought the fruit would rot. I said I would guarantee his present tonnage per acre from the harvesting he had done. I also told him there were going to be some angry winemakers (even though their own fault) when they discovered they had battery acid. He did irrigate as I instructed – sprinklers on at 9:00 AM and off at 3:30 PM allowing the vines to completely dry before night dew appeared. Two weeks later we harvested beautiful fruit at proper numbers and ripeness.

After a good rain or irrigation there *will* be the *taste* of water in the grapes. Depending upon the quantity of water applied and the subsequent atmospheric conditions that taste of water will go away – in about 2 to 4 days. *Then*, when the numbers are as one desires for the proposed type of wine, the grapes can be gathered. There will be *no* affect upon the finished wine. Winemakers often add water to grapes or juice pre-fermentation and the magic of fermentation erases any effects with that practice. Why would one think it would be any different in the field? In fact, it isn't. The only thing different is *who* adds the water – the winery or the grower. I might also point out that at the winery the law was "water incidental to processing is allowed" but not deliberate water to reduce sugar (though it was done). Water in the vineyard had no such restriction. That law has recently been changed allowing wineries to add water at will leading to other abuses.

During the growing year of 1976 we began a procedure we called "TUCKING" – by hand positioning the growing shoots to the north (upwind) side of the second wire and, later on, of the third wire. We were also careful to leave one or two shoots to the south side of the second wire and, later on, tuck these to the north side of the third. This practice was to prevent the young cordon arm from rolling over to the south (downwind) side thus to force the shoots upright. Failure to do this would often cause the cordon arm to snap. It also correctly positioned the future spurs upright. One can easily see from the expense of this that the desire for a more efficient method arose. Later on the movable trap wire of Germany and Alsace came from this desire and solved both the expense and south wind situation. There were eventually many who copied the step without leaving the south side shoots. In our area when a good storm is coming the wind shifts to the south. If one has not left those southside shoots the entire canopy will roll to the north snapping cordons and requiring repositioning of the canopy – usually when well developed and therefore hard work (expensive). The trap wire concept precludes this by trapping the shoots between the permanent wires and the movable wires. It also exposes basal leaves to light and the fruit to diffused light.

1977 was the first small commercial harvest for Ventana. In addition to a variety of home winemakers and small Santa Cruz mountain wineries we supplied grapes to Merry Edwards at Mt. Eden and Ahlgren Vineyards. We worked with them on the selective harvest approach and the Ahlgren Vineyards 1977 "VENTANA" Chardonnay received a gold medal at the Orange County Fair – perceived as the most prestigious at the time because the judges were winemakers and/or winery principals. We supplied Riesling to Stony

Ridge Winery for a late harvest (Botrysized) wine that also received a gold medal.

Botrytis, as I have mentioned, was a significant problem at the time. To the big wineries "rot was rot" and they had no tolerance for botrytis. They really had little experience with it in the Central Valley as a stand alone. America was still in a somewhat sweet mode and I found Botrytis rather fascinating even though not desired and fought in the field. Amateur winemaking was, and is, a big thing in California. There are many, many skilled home winemakers here and I owe a lot to them. In those days the California State Fair wine judging was "Amateur Only" – there was no commercial wine judging. The only wines eligible had to have received medals at a county fair judging. They awarded one Gold Medal only for the year for the State of California. I received it in 1978 for a 1977 late harvest Johannesburg Riesling. All of our studies through harvest 1977 had been conducted under the federal home winemaker rules of 200 gallons per year per head of household. I and a few of my guys had duly registered ourselves. Starting in 1976 I supplied grapes to home winemakers and we continued that up through 2006. Part of the sales price was that when bottled they had to give me one bottle of each wine the following harvest when they came. After harvest I and our team would gather together and taste and analyze each wine. I made many phone calls asking people to take me step by detailed step of everything that they did from the time I gave them the grapes or, later on, the juice. For some wines I wanted to know how they had made such a lovely wine. In other cases I was trying to learn what NOT to do (though I didn't say that)! The same approach was followed with small bonded wineries although in those cases I would travel to the wineries at various times, taste and sort of "hang out" talking wine, buying some wines and just generally make a pest of myself asking questions.

Some years later I recalled this process while lecturing to a group that included a few of said small-winery winemakers. One of whom piped up saying "Yah – but we didn't know the

SOB had a photographic memory or what he was doing". Everyone had a good chuckle.

The "Monterey Veggie" reputation had taken root and, in fact, in many cases of commercial wines it was warranted. Even in home wines. Reds were certainly problem children – *particularly* so the Cabernet family. But white wines were another matter entirely. When I mentioned earlier that my initial wines were horrible I meant from oxidation and lousy winemaking or storage on my part – not from "veggies". In fact the early aromas of those white wines were lovely. The interesting question was how could some wines of the same variety grown, harvested and processed by one entity be so vegetal and those by another not be – all in the same climate?

At that time we were using the selective picking on Chardonnay. It diminished the problem but in some cases did not completely eliminate it. Upon close observation it was apparent that - trying as hard as they could – the crew was unavoidably getting some less ripe fruit into the mix. Some color distinctions were too close. We had to perfect a system of growing that would bring all the fruit into the zone at the same time. The *theoretical* basis was in mind but the system was not yet in place – the need was very apparent.

An interesting variety was Riesling. Riesling displayed absolutely none of the veggie components – not in its developmental precursor stages nor in its "ripened" stages. However, I observed it in some commercial wines. That confused me for awhile until I discovered in my prowling around that in those instances there had been some "blending" off of cold climate Wente clone Sauvignon Blanc (*not* by Wente Bros Winery!) into the Riesling wine. I have seen many levels of farming – good to horrible – and know of no way to induce veggies in Riesling or Gewurztraminer in the field. In the winery – yes, by blending in something else but not in the fruit itself. Those problems were winery caused – not vineyard. The media could not see this so damned the region.

That year – 1977 – we made some wine from Pinot Noir and from Petite Sirah grapes. We de-stemmed and fermented in open top bins. Everything went well and we pressed off the "dry" wine which measured dry *before* we pressed. I did not check it *after* pressing! We barreled down after a cursory settling and – because of my fear of oxygen – we drove the bungs home. The next morning we came into the barn and there was red all over the place. I looked to the top of the barn and there was a red circle only about 18 inches in diameter. What the hell was going on? Then I looked more clearly at the remaining barrels. The ends were bowed out like balloons! I moved all my people back out of the way, took a long 2 x 4 and proceeded to knock the bungs loose from these wooden "bombs". It was exciting. It was also a huge mess. I had now learned about sugar in wine being released by pressing, about pressure in a closed container and more things not to do. As it turned out I was destined to learn about a lot of things NOT TO DO over the next couple of decades. Sometimes those are the best things to know. They can be expensive lessons, though.

The summary of that farming year was that we leaf pulled, shoot tucked, selective harvested to color and farmed until we had desirable numbers although still higher in pH and TA than we really would like. Our activities of exploring other Salinas Valley properties took place after others had harvested or abandoned fields. I was interested in the different flavours generated in different vineyards in different parts of the Valley and different soils. We were also able – by hand selective picking – to supply reasonable quality fruit according to the standards of the time to other wineries and winemakers. In each case they knew the fruit was not from Ventana because I usually took them to the locations to taste the fruit before harvest. Those from "The Ventana" were so identified.

By harvest year 1978 the problems for Monterey fruit were becoming pronounced and financial difficulties were growing for the cool climate industry. It was imperative that we find solutions. Already bankruptcies had occurred and some vineyards had been foreclosed upon by insurance companies and banks. Some foreclosed vineyards were "mothballed"- essentially just not farmed. Others were torn out and returned to row-cropping. More would follow over the next decade or so. The money people were bailing out trying their best to salvage something from the disaster or just walking away absorbing their losses and looking for the next game. Sadly for me, I guess, I was by this time a believer in the potential for greatness for Monterey! It was not a fanatical and mystical "belief" rather it was a belief based upon what I construed as a sound analysis of the circumstances. Unfortunately, it would take many years and much money for experimentation and testing if the hypothesis was to rise to the level of a theory. Fortunately, I had a lot of years ahead of me of the time part. Unfortunately, I had very little of the money part. My very public articulation of my blasphemous and heretical ideas (all the time) so in dissonance with the conventional academic wisdom helped to drive away any potential outside monetary assistance. So be it – faint heart never won fair maiden!

In 1978 I bonded the winery as a vehicle to test the viticultural methods and to be able to sell the results therefrom. It appeared to me that I could speak about the ideas until hell froze over and it would have no effect whatsoever against the entrenched mindset. After all – who the hell was I – just a Navy attack fighter pilot from Washington State whose last name did not end in a vowel! One person then referred to me as a "distasteful barbarian warrior in Levis speaking garbage". Flattery gets to me every time. It had already become clear to me that different winemaking procedures would result in different wines. Pretty astounding observation – no? Sometimes I amaze myself. I needed to have control of those procedures in order to null-out the winemaking variables (as much as possible) such that the finished wines would demonstrate the differences in farming procedures. I also needed to explore winemaking procedures that would aid our region as an interim until I could get the new viticultural system up and running. These procedures would have to be geared to cool-climate fruit. The extant procedures common in the industry had evolved for warm and hot climate grapes. In addition, I needed that enology knowledge as a "feedback" to farming practices – a foreign approach at the time in America. My intent was to solve any winemaking problems in the vineyard as much as possible and to find winemaking procedures to handle those problems that we couldn't.

The work really accelerated that harvest year. Low and no SO2 usage was one exploratory project. Multiple yeast strain usage was another, Malo-lactic fermentation of whites another. Flavour comparisons between same variety grapes from different parts of the valley was another. Comparative barrel effects was another. Differences between grapes grown in different vineyard configurations was another. Busy, busy, busy.

In *most* cases then and *many* cases today when one opens a bottle of white wine one gets a heavy dose of sulphur dioxide. There are people who are physically intolerant of that. However, I am not one. I just simply didn't like it at all – in fact, I found it repugnant and off-putting. When I looked into the recommended practices I found much to dislike and with which to disagree.

The standard recommendations and practice was to add 50 ppm potassium metabisulphite powder to the *immaculate grapes* prior to (or at) the crusher. If there was *any* sign of rot or deterioration increase the dosage to 100 ppm. The larger wineries would require the grower to add coffee cans of potassium metabisulphite powder to the bottom of truck tubs prior to dumping fruit into them. Some would require dusting the powder over the top. Those grapes sat in those trucks often up to 24 hours traveling long distances then waiting in lines to dump! In 1978 I followed those 50 ppm procedures for most of our grapes but I had serious questions about the need for that much. For some small quantities of grapes I reduced the quantity to 25 ppm but still followed the normal procedures of adding it at the crusher before the press (white grapes). At that time I had a Howard rotopress – a plate press. The same procedure was followed for reds – add the sulphite at crusher before de-stemming prior to transfer to fermentors.

We were essentially a barrel fermentation operation for Chardonnay so it was very easy to use many different yeast cultures on that variety. Some yeasts were commercial strains. Some were supplied to me by Lisa Van de Water's Wine Lab. Others were from small wineries as "proprietary" strains – supposedly. The differences were astounding. Lisa was an early proponent of different yeast strains for different purposes but she, too, was only listened to by some small wineries and home winemakers.

The "method" of adding yeast was also questioned by me. U.C. Davis recommended *large* "starters" added to juice to get the fermentation going fast. That is probably a good procedure for industrial wine in terms of speed of turn-around time of tanks and production line processing to wine. I had come to the idea that a slow start was more beneficial to the finished wine. Glycerol in a wine adds a sense of "thickness" to wine as well as a hint of sweetness. The human perceives glycerol as sweet. Just as glycerol formation is a by-product of M-L fermentation it is also formed in small quantities by yeast cell division until the density of yeast cells reaches the fermentation point. A small starter elongates that period allowing the formation of more glycerol and thus more "mouth feel" in my opinion.

On reds we would add the yeast in one corner of the vessel letting it slowly spread across the mass. For whites, after settling and chilling juice for 36 hours, we would rack the clear juice to tank or barrel and then add the yeast to the top.

One exception to this procedure for white wines is Chardonnay. After experimentation and the guidance of Dick Graff of Chalone we would rack the clear juice to barrel but not as full as desired. We would then work our way down into the settled "gunk" in the settling tank distributing the fluffy stuff evenly back in the barrels to proper fill level. Chardonnay and Pinot Blanc are the two white varieties in which we want some of the "fluff". It adds body, aids full fermentation and adds flavours. This must not be done with Riesling and Gewurz in my opinion.

The Malo-lactic work continued and I began focusing upon perceived pH limitations accepted by the industry – limitations I just couldn't believe. I thought that upper Monterey should be ideal champagne country given its climate comparisons with France's champagne area. I had rather liked champagne since 1972. Mirassou made small quantities of various champagnes and Ed Mirassou allowed me to have a few cases a year of their "Natural" – a completely dry bubbly. It would quench thirst. I had purchased many French champagnes over the years and run them through analysis. I knew that all French champagnes had undergone M-L fermentation. I also knew from the French literature that the base wines would have pHs as low as 2.6 or 2.8. Yet, it was America's belief that M-L could only be conducted down to 3.2 and below that it was *impossible*? Proof? Go to old copies of the ASE (American Society of Enology – it would be years before Viticulture was added) journal. You will see ads for an M-L strain of bacteria called PSU-1 (Penn State University -1) advertised as *THE* low pH strain – good down to 3.2 pH! We brought in small amounts of grapes at low pH and, sure enough, I couldn't get them to undergo the M-L fermentation. That problem needed more work.

I continued mapping the flavour and compositional structure of the Valley on Chardonnay – north and south, east side versus westside! The barrel fermentation approach allowed small lot comparisons although we were aware of barrel differences not due to location and tried to guard against it. Our small scale did not give us "proofs" but it did give us a "feel" for differences. The numbers, of course, spoke to dependable differences.

The barrel studies continued utilizing French barrels from Sirugue through Jim Boswell in San Francisco. These were the same barrels used by Dick Graff at Chalone and Mt. Eden. Dick had recommended them to me and I had become more familiar with them when Merry Edwards bought fruit from me in 1977. I also had some new Blue Grass Cooperage American oak barrels. Time would tell.

In 1978 an acquaintance who was in the business of manufacturing stainless chambers holding membrane cartridges or tubes for filtering usage in the chemical and medical areas spoke to me about potential applications in sterile filtration of wines. The usual unit then was an in-line holder. I commissioned him to build me a unit that would hold seven membrane cartridges. This technology opened a whole new panorama for us in winemaking. The ability to filter down to .45 microns would give stability without chemical sterilants or other techniques. By "chemical" sterilants I mean the various ways commercial wineries tried for stability – the sorbates (geranium smell), heavy use of sulfur, the "blue" fining of Germany, etc., etc. I had read of those but never used any of those approaches. The new filtering obviated any need (if any existed). I was philosophically against using any such chemicals – which was why I went to lengths to embrace this new tool. I don't know for sure if the technology was in use by any of the older wineries yet but I suspect not as the unit was experimental to me and commissioned by me to be built. It is, of course, possible that some other manufacturer was working with another winery outside of my knowledge but I have not heard of it, if so. Shortly thereafter Harry Rosingana at Stony Ridge Winery acquired a unit. If my memory serves, I think it was Harry that first introduced me to the fellow who sold me on the idea.

Our basic approach in 1978 was pretty straightforward. The Chardonnay received its 25 ppm at the crusher (a reduction of 25 ppm from U.C. Davis' 50 ppm recommendation), the juice chilled and settled then transferred to barrel for fermentation. When dry it received 35 ppm sulphite and allowed to settle for a month then racked using CO2 gas pressure. The wine was *fined* with egg whites, settled and racked again all the time maintaining 33 – 35 ppm and eventually sterile filtered and bottled. The Riesling was stainless fermented and alternately stopped and started following the German nitrogen depletion approach until it wouldn't restart then settled and filtered again maintaining 33 – 35 ppm until bottling. Both wines went on to receive multiple gold medals at major judgings.

In 1979 we made a lot of progress in understanding the interworkings between vineyard and winery. We also experienced major breakthroughs. Early in the year a previous summer intern from Fresno State who had graduated and gone to work for Seitz – Paul Smith – gave me a theoretical Seitz paper on no SO2 utilization. Paul knew of my interest in lower SO2 use. What a revelation! That German paper led me to explore their assertions and led my mind into far reaching areas.

Project A

I was (and am) of the opinion that the coolness of the upper Monterey is better suited for quality champagne type wine made from Pinot Noir and Chardonnay than grapes from warmer areas of the state. At that time the Ventana had 16 acres of Zinfandel. It was a problem child and was soon changed to Chardonnay – I believe Zinfandel should not be grown in cool climates. That variety is notorious for not ripening evenly across the vine *and* across the cluster. It needs warmer climate to minimize those variations. In '79 I decided to use zinfandel for base wine for champagne cuvee for several reasons. I was receiving $800/ton for Chard and Pinot and zero dollars for Zin. I needed to see if I could get *red* grapes from the vine procedurally through the press and get reasonably light colored juice. If I could do it with Zin then Pinot would be no problem.

Finally, I could play with this valueless juice on the Seitz ideas.

We successfully pressed the Zinfandel to reasonable light juice that ultimately had only the faintest tinge of pink tone. We barreled it down for fermentation using no SO2 at any point. The pH was 2.8. We watched the predicted "browning" and at the end it did drop out as predicted by Seitz. Elsewhere in the winery we had Malo-lactic fermentations going on. I had no idea that M-L bacteria are as virulent as they are. Along the way that juice went through M-L! Wow! At a full four tenths lower than the believed minimum pH in a M-L resistant variety at cool temperatures and in accord with champagne procedures! Double Wow!! This was a major event to me.

Project B

We took a small amount of Chardonnay (about 2 ½ tons) at sugar and pressed to a stainless vessel with no SO2. We racked to another, inoculated with yeast and watched the browning occur. At around half the yeast fermentation we added M-L bacteria. We had already learned over the previous several years of experiments that M-L fermentations could go awry particularly if the bacteria could get to sugar once *fermentation* began. By adding bacteria at about half yeast fermentation we had found that there is enough heat for bacteria expansion, plenty of nutrients and yet the sugar fermentation will complete before the M-2 bacteria density is sufficient to seek the sugar. We were finishing M-L within about thirty days of primary. The predicted dropping out of the brownish tinges was as the Seitz paper said but with a little deeper golden color. We found that a little Polyclar AT took care of that (AT is, simplified, ground up nylon which drops out or is filtered out. It does not stay in the wine).

We also noted that the wine from this procedure had *none* of the residual touch of bitterness common to Chardonnay wines if dry as many still have today!

Project C

We purchased a small amount of Muscat Blanc from a nearby vineyard. Another small winery also had purchased some. We picked at the same time – side by side. Their finished wine had to go to about 5% residual sugar to cover the bitterness. Ours was covered at 1%. The differences were twofold: 1) we didn't press as hard and 2) we added SO2 to the pressed *juice not to the grapes at the crusher*. The SO2 at crusher extracts some bitter component from the skins and seeds.

Recall that earlier I commented on SO2 in the trucks by the big wineries and the duration of the skins on that SO2. I am of the opinion that SO2 addition at any time before the juice pan of the press is detrimental to the subsequent wine. If needed, addition at the juice stage requires far less for any needed purpose. Further, total SO2 in the finished wine will be substantially less. Less total SO2 reduces what I call "SOAPINESS" in a finished wine when too much is present.

Project D

As mentioned earlier, the variety U.C. Davis carried as "Savagnin Musque" then later re-named as Sauvignon Musque and, later, as I believed – Sauvignon Blanc – was made into wine in glass carboys. Another winery also made glass carboys of the same grapes. Later, at a technical group meeting, both wines were tasted. Ours was a gorgeous little wine that displayed the future desirability of this grape. The other wine was very vegetal. I do not know their procedures but it didn't matter – all the group could lock on was the cliché vegetal aspect and that added some years to its recognition and acceptance. Thanks to Terrell West's perseverance it has reached its present stature. There were a few years when I was so busy during harvest and Terrell would come and force me to continue the wines each year.

A little background history is needed here for refresher even though I have written and been often quoted on this matter elsewhere. When I came into this industry and observed field "chip budding" (the industry standard) I was amazed at its slowness and cost. Steve Alderete was the viticulturist for a large local operation and he had a lifetime of practical experience. I asked him why they didn't use "T budding" as was used in apples – it was much faster. Steve thought about it and commented that he thought he remembered his father saying something about it or doing it. Otherwise he had never heard of it used in grapes. He, Terrell West and I played around with grape bark to see if there was enough slippage to be used. There was. Steven and Terrell became very interested in the procedure while I was off in other things. They involved Curt Alley of U.C. Davis who became very active in the project. Curt's early papers on it had all three names at the top. Terrell donated the land and farming for the field trials. Terrell told Curt that as long as they were going to test field T budding they should also do varietal testing in the process. Terrell asked Curt to select off-the-wall weird varieties from Davis' collection that he thought should be looked at in our cool windy situation. Curt did. Boy, were they off-the-wall- to me, then. I'd never heard of these things nor had Terrell or Steve. The T-budding was successful and became a widely used tool in vineyard change-over.

The tool was needed because during the big planting boom world (and U.S.) consumption was roughly 65% red, 10% pink and 25% white and vineyards were planted that way. With stainless steel tanks, temperature controlled fermentation and gas protections white wines were now gorgeous rascals. The public reversed themselves and the vineyards were thus backwards. Also, certain reds in Monterey were horrible then. Changing variety in the field on older vines was viewed as necessary.

In my case I loved Sauvignon Blanc – both from Europe and from Wente Brothers. There was only one clone extant in California – the "Wente Clone". It made nice wines in warmer climes but from cold climate under conventional vineyard design it was horribly vegetal – often like canned asparagus. It could be manipulated somewhat but not adequately. What was grown in Monterey was mostly used to blend into water – like central valley Thompson wine.

Sauvignon was my love and I couldn't grow it here. Yet, I knew that Sauvignon grew in areas of France that had coolness like ours without traces of veggie. Areas of Poitou, for example, handled it yet, though, once-in-awhile one could find a bit of it. There had to be clones amenable to our situation and I made it a future project to find one. At that time my plate was full.

That year – as the fruit ripened – Terrell came to me and wanted me to walk his test block and taste fruit. The block was 5 vines of each of the varieties with 5 repetitions. The vines were number coded in the field. Each time we came to this one variety I reacted, said this is Sauvignon and I need to make wine from it. Each time Terrell said there is no Sauvignon Blanc out here and to identify this variety we needed to go to the code master in the office. We did. He was right – no Sauvignon Blanc. It was something Davis called "SAVAGNIN MUSQUE". I was positive it was Sauvignon. No matter, we made a test wine as noted above.

The next year I was so swamped with study projects and business that the Sauvignon project slipped my mind. Terrell showed up at my winery. He reminded me that the Sauvignon would soon be ready. I begged off saying I simply didn't have time this year. He said "Godamn it, I'm going to pick these grapes when they're ready, haul them down here and YOU are damn-well going to make that wine!" I'd never heard him talk that way or even using that many words at once. Those who know Terrell will understand what I mean. We looked at each other for a bit then I said "Oh –o.k. When are we going to do this?". We made wine. It was gorgeous – all four barrels of it. We labeled it Sauvignon Musque.

In 1981, and again in 1984, Professor Pierre Galet of Montpelier University visited the U.S. He positively identified the vine as Sauvignon both times under controlled conditions – both in my hands and in the hands of our county's viticulture extension agent.

This clone is of major importance in California's plant material library and has made substantial contribution to our continually advancing wine quality. I am proud to have been part of that and for the opportunity to grow my favorite white wine here.

In more recent years when Mondavi (and others) were contracting with growers to plant Sauvignon and requiring it to be this clone my phone was constantly ringing with growers wanting to know about this clone that was being required. I answered all their questions (over and over) and faxed a copy of a 1988 writing I had done (included in the appendix).

Today, this clone is of tremendous importance to our industry and has allowed Monterey to consistently produce world-class Sauvignon.

Project E

Pinot Noir – ah, Pinot Noir – the bete noir of so many, myself included. The young vines and our lack of knowledge on how to grow them properly limited our exploratory work. We generated a nice little red pinot noir wine without the "burnt rubber" smell. In fact, the wine had a nice little floral touch.

The grapes were hand picked, de-stemmed to an open top fermentor. We used 25 ppm SO2 at the de-stemmer. Twenty four hours later we inoculated with yeast in a corner of the fermentor. When about half done we added bacteria. We punched down manually several times a day. We pressed at dry indication and finished fermenting and M-L in tank and then to barrel.

This was the year that I had bought a full container of Sirugue barrels – 150 of them. According to Jim Boswell, that was the first time one winery had purchased a full container themselves. To the best of his knowledge other wineries had purchased small quantities.

There is a funny little story behind this. Dick Graff of Chalone and Mt. Eden had introduced me to this barrel. In 1978 I had bought a few when Dick let me tag onto his order. Recall that Dick and I had formed a small technical group together, shared knowledge and worked closely together. In 1979 I called Jim Boswell (Sirugue's barrel agent here) and tried to order some barrels. I couldn't get them until December when his next container came in. I needed barrels by the end of September – October at the latest. He couldn't do it. I asked "What if I bought a full container, could Yves Sirugue supply and ship now?". Jim choked, said he'd check but was sure he could. It just hadn't been done before. Soon Jim called back and we had a deal. What I didn't know was for that big a deal he had put me in first place and was sliding everyone else back. Later, when Dick was informed he wouldn't get his barrels until later and the reason therefore – that I had bought Sirugue's current supply – he was irate and felt betrayed. He called me and was furious – beyond furious even. I listened. When he paused for breath I asked him how many barrels he had ordered. He spit out the word "five". I said "Dick, I didn't know any of this. I have no problem with you taking 5 from my container and give me your 5 whenever they come in. Not a problem". Everyone was happy – problem solved.

Sirugue then was my preferred barrel though subsequently I also found great favor with Francoise Freres which I also used heavily through the eighties and still do. I also purchased some Seguin-Moreau and Taransaud barrels for use with Bordeaux varieties, though now I prefer Tonnellerie Bordelaise. The Sirugue and Francois Freres barrels were geared to Chardonnay, Pinot Noir, Chenin Blanc and Pinot Blanc

Project F

While engaging in our business of buying and selling grapes I found one interesting little block north of Ventana a few miles and on the east side of the valley. It was a little point of vineyard of about 1 acre or so and part of a very big vineyard that had been planted in '72 or '73. The plant material had come from a central valley nursery as dormant rootings but no one was quite sure which nursery. I found the flavours rather interesting and so I made wine from them. I thought they were Sauvignon and the planting chart of the vineyard showed Sauvignon but they tasted different from the usual Wente clone from our area. That following winter the grower tore out that little 1 acre of vines and planted it to the adjacent variety for efficiency so the plant material – whatever it was – is lost to us forever. The blasting of Monterey for "veggies" was in full force in the marketplace and it was not a pleasant time. When this wine received a Gold Medal at a major judging one could almost hear a unified sigh of relief. It went on to receive other golds as well so it wasn't a "fluke".

My focus upon Sauvignon is because I love it as a table wine. It is such a food enhancer with its typical crispness. In wandering through Europe I will generally order one if available although a well-made burgundy also will work nicely as well as a dry Riesling. There are two general

Sauvignon tablewine styles (ignoring the temporary – I hope - American thing of the sweetened version): the Loire style and the Bordeaux. The Loire style of Sancerre, Pouilly Sur Loire (in the form of Blanc Fume) and Menetou-Salon are essentially straight Sauvignon from inland areas along the Loire River fermented dry.

The cooler Bordeaux region Sauvignons were of more professional interest to me. In the French areas the wines didn't go by varietal name rather by the estate name. One simply had to "know". In Bordeaux the whites were/are blends of Sauvignon and Semillon (ignoring Muscadelle) – in most cases heavier to Semillon. A common saying in the past (and probably still) was "Sauvignon is for the youth of the wine. Semillon is for the aging of the wine and to control the savageness or wildness (the French use one word "SAUVAGE") of the Sauvignon". Then, the Semillon of Monterey was almost as aggressive as Sauvignon. So – considering the asserted *function* of the Semillon in the bordellaise mind I went to Chardonnay which has similar characteristics in that they both develop lusciousness with time in bottle. Chard's neutrality diminishes Sauvignon's assertiveness. That '79 had about 15% Chardonnay in it. Gorgeous wine. Ever since my Sauvignons have had around 10% Chardonnay. I have shown that trick to many winemakers over the years. I do *not* M-L Sauvignon. I do like to barrel ferment a portion in older more neutral barrels for the smoothing effect and stainless ferment the other portion for crispiness and fruitiness. I, artistically, do not wish the taste of oak to show much in Sauvignon. Hopefully, the Chardonnay – wannabe Sauvignons are history – however some heavy-on-oak Sauvignons are produced today in Bordeaux.

Project G

The Chardonnay production was pretty straightforward except that we were using 12 different yeast strains in comparative trials in barrel fermentation. We were also doing verification trials on my opinion that SO_2 added to grapes before pressing extracted bitterness that was apparent in the finished wine. This was the last year we added any SO_2 to white grapes prior to pressing. Since then it is always added to juice when used. We were also conducting taste and pH of juice taken at different press pressures. Hard squeezed juice had very high pH, was oxidized looking and tasted lousy. We stopped being greedy at the press. Later, when I convinced some winemakers that they could get a much better product and save freight money (then wineries paid freight) if they let us press to juice and they tanked it in. No waiting at jammed up pressing stations, pressing as the grapes left the field, field heat reduced quickly, held and shipped under CO_2 blanket, etcetera. Of course, initially there was resistance about such a heretical and radical idea – my god, let a grower press to juice? I always offered to each client to have someone present while pressing. There were always discussions about gallons per ton and dollars per gallon. I had such difficulty trying to explain about greedy pressing and I think most thought I was trying to chisel down the gallons per ton yield so I could keep the extra. I gave up and simply absorbed the difference actually providing more gallons per ton than I actually got. Much of the time most buyers never showed up at pressing time. Over the years I only had one fellow from Napa who showed up every year at pressing time of his Sauvignon. The first year as we were pressing I said "all right – that's a cut". He said "Hey – there's still juice coming". I took a lab beaker, caught some juice and said "taste it and then put in on the pH machine. Then tell me if you want that in your wine. If so I'll keep pressing until the wind carries away the residue". He took one taste and told me to stop. He couldn't believe the pH meter reading. He was a very skilled red wine man but he learned a bit about whites that day. We had a long and fine relationship.

What I look for in a wine is a "harmonious balance" of qualities based upon six criteria ruled by one dictum. They are: color, aroma of type, taste, texture, finish and absence of bitterness or other off-putting flaws. The over-riding dictum is: is the wine compatible with food – that is, is it food-friendly? These criteria are primarily a "teaching guide" or starting base for analysis. Experience and/or an innate mysterious artistic perception allows development of a "feel" for the "rhythm" of the growing season. The wine master "senses" the rhythm of the fermentation and the elevation of the wine. The constant question (daily) in the cellar is – is the creative process unfolding as it should? Basically, a wine should look good, smell good and taste good. It really is that simple. Getting there is often the difficult part!

There are a myriad of characteristics in wine each of which brings a frown to my face and do not give me pleasure (after all, that is the main modern-day purpose of wine). High alcohol I find to be very off-putting – both for professional reasons and personal. I have never acquired an affection for going to the glass and getting a blow-torch assault on my nasal passages. It is a personal quirk. I find it unpleasant to say the least. During the peak of the "fad" of high alcohol I saw some wines that raised definite worries about open flame at the table. I do not comprehend the appreciation (or tolerance?) by some for this aspect. Once I had the temerity to recommend trying whisky for even more "pleasurable" nose – if that is what it was.

Professionally, I note that levels of wine alcohol are primarily a function of the sugar level of the harvested grape. High alcohol means a high sugar level. Unfortunately, as sugar accumulates in a grape that is not the only thing that is happening. The other compounds in the grape are also changing. Delicacy and subtleness are lost in over-ripe fruit. Most over-ripe red grapes develop a prune juice component and/or a raisin effect that carries forward into the wine. These are tastes that I do not appreciate. I may be an old geezer but I do not yet need prune juice. Besides – I don't like it. However, some folks do, I guess.

Another problem with over-ripe fruit is that the pH climbs, often drastically, to nearly unworkable levels. I have had winemakers (small operations) tell me that they just don't worry about it. To me, pH of the juice and pH of the finished wine are critical factors. In my opinion, high pH wines simply will not age properly. In fact, I believe they have no chance whatsoever of properly evolving in bottle. If high pH wines are to one's liking *and* one intends to consume them soon – fine. But don't kid yourself about "putting them down". You'll be disappointed. Also, the higher the pH the more susceptible the wine is to bacterial infections. To have a winery full of high pH wines is to create an ideal situation for breeding of these organisms. Once infected it is very difficult to "clean up" a winery.

Once, a long time ago in Salt Lake City, I was the invitee at a private home wine evaluation gathering. (Utah has some bizarre laws). In the course of my talk I spoke on the above points adding that those who had "laid down" high alcohol Pinot Noirs (quite the rage then) would be very disappointed. I said all that would be left was alcohol and wood with no fruit or other merits. The husband of my hostess, as it turned out, was one of those and he took immediate umbrage. After some strong discussion be raced to his cellar and brought up several of his treasures to share with all and to put me in my place as a charlatan speaking with forked tongue. Do I have to tell you the results? Needless to say, that was one chagrined hombre!

The high alcohol "fad" seems to be waning – happily. This was a gross misdirection for our industry (not to mention possibly socially irresponsible) that we are weaning ourselves from – primarily by many skilled winemakers finally fighting back. This "bigger is better" nonsense came about, I think, by scribes and pundits using words like "huge", "gobs", "monster", "powerful" in conjunction with praise and scores.

Absent were words like "graceful", "balanced", delightful", "enticing", etc. – probably because it takes some knowledge or developed taste to discern those elements. Egotistical owners (and some winemakers themselves) got caught up in the hunt for scores from the Parkers and Laubes of the world and pushed their winemakers to make these sorts of wines. It just grew and grew, the wines getting bigger and bigger and their true merit tumbled. Many wines became caricatures of wine but sadly still received raves and high scores from the sellers of magazines and newsletters. Grace, elegance and subtly were temporarily cast upon the junk pile as was ageability and pleasure.

To compound the problem it was "discovered" by many that the scribes *really* loved oak – the more the better! In many cases of high scoring wines (read Chardonnay) it appears that the sole purpose of the wine was to act as a liquid carrier of oak extract! I am serious! I bought a few of these wines to explore what the scribes and critics were raving about and I was aghast. I could find no indications of the varietal in the wine – all one could smell or taste was oak. If these commentators were at all correct then a strong logical case could be made that very high quality Chardonnay (and why even use Chard?) could be grown in a Region 5 desert! Incredible. I saw this trend starting and began to publicly speak against it in 1981. Here we are 27 years later and we finally see the reversal underway. I probably contributed somewhat to the oak thing. I was, as you know, deep into the exploration thing. In 1978 and 1979 I had mostly new French oak barrels (discussed elsewhere). The Ventana Chardonnays of those years were the "darlings" of the street. Some years ago The Wine Spectator did a piece on Cecil de Loach of De Loach Vineyards wherein he stated that when he tasted the 1978 Ventana Chardonnay he *had* to make that wine. Not too many years ago Dan Lee of Morgan Winery asserted that the 1979 Ventana Chardonnay was the best ever made in California. To my palate they were experimental wines and over – oaked as necessity. It was all I had. Seeing this copy trend developing, I spoke long and hard against it – to no avail.

Don't get me wrong – oak *is* needed and desirable in certain wines but the oak should be the frame around the picture – not the picture itself! I think wine should taste of fruit not resin. Of course, I may be biased because I grow grapes not barrels.

Well, actually that may not be completely true. In 1978 I planted about 1500 French barrel oak trees on the Ventana. They are still mostly there standing about 60 feet tall. So far there have been no barrels made from them. My intent was to grow them to sufficient size, harvest a few, air dry them, grind them up and make an extract. At the same time I intended to get comparable aged pieces of wood from a barrel-maker friend in France and do the same with it. Then we could compare to find out if the tree produced the same resins here as in Europe. That work is still to be done. However, I can tell you that the tree itself thrives here but with irrigation. In another 50 or 100 years we should have some good information in this area.

I repeatedly speak of "Art" and "artistry" in the wine context. In evaluating a wine I believe that one should separate technical competency from artistic. Technical mistakes or successes are a function of knowledge and diligent application thereof! Technical deficiencies in a wine are recognizable. Art and artistry are harder to define but are like obscenity – I can't define it but I know it when I see it (quoting a Supreme Court Justice). Roughly, I would call it the "style" and its "presentation" meaning the balance of the factors. One can easily critique technical competency but one should refrain from critiquing artistry beyond "I like it" – "I don't like it". Art doesn't lend itself to anything other than personal perception.

Other Projects

In the beginning period of the big planting boom of the early to mid seventies hubris was rampant. Money was flowing in from investors and institutions by the trainloads. The promoters had tied in the imprimatur of the U.C. Davis academics so they had the illusion of possession of ultimate and eternal truths working for them. At times I felt sorry for these academics as one professor after another was trotted out before fawning city folk to cite absolutes. They were being hustled beyond belief by tax-shelter, smoke and mirror manipulators – and were totally unaware. These were not experienced worldly academics rather they were quiet provincial students of a backward discipline locked in tradition in a Podunk university ignored by most of American culture and, historically, even negatively regarded by many mainstream channels. Strong? You bet! But *that was* the nature of America before 1970. Wine was for winos – derelict alcoholics – and certain groups looked down upon by many – Catholics and Italians. This was the world these academics had historically served and there certainly was no prestige in that outside their own tiny social, bucolic culture. Then, around the sixties, America commenced a huge cultural shift. It probably, started, seminally, with James Beard's Cookbook (1959) but the ground swell was caused by "Mastering the Art of French Cooking" by Julia Child and Simone Beck (1961). Child's TV show "The French Chef" (1963 – WGBH public TV – Boston) was a raving success rebroadcast by many stations across the nation. Also in 1963 Craig Claiborne's regular column of restaurant reviews hit the New York Times. It became "in" to talk and cook foods. Once foods become important so do wines. Once a person learns the difference between cilantro and parsley one can appreciate the difference between Riesling and Chardonnay! Of course, in 1977 Alice Waters at Chez Panisse brought the act to the west coast throwing more fuel to the growing fire.

The stage for the seventies was set by the east-coasters of the sixties.

From the womb of anonymity these provincial academics were wined, dined, fawned upon and drug to the podiums in front of many movers and shakers. The prestige was intoxicating! They were sitting ducks. I often watched as an academic in these situations would *attempt* to put in all the academic caveats, ifs and buts on a subject and then be cornered and pushed into an absolute assertion.

The promoters, themselves skilled at "promoting", were totally ignorant of the ins and outs of the industry they were promoting – essentially all were "outsiders". However, armed with academic and financial support they charged ahead. I charged with them. Little did we know we were The Light Brigade charging into disaster. I am not exempt from the above indictment. I was an investor and was pulled in. I took in all the professors had to say as gospel, initially. I began here in the spring of 1972 after I finished my Navy time but I was an initial investor founder since 1969 in the company that brought me here. By late 1972 I realized that all was not well in Eden! By 1974 I realized that there were horrendous problems to be solved, a lifetime and more of work to be done (to any young people reading this – don't worry, I've left more than enough for you to solve and do!).

Previously I have discussed various viticultural and enological aspects of the new viticulture which will be summarized later. While that whole evaluation was occurring there were other, associated projects underway. I will address each in no particular chronological order or pattern as they were all parallel to varying degrees. I have discussed the finding and elevation of the Sauvignon Musque clone in other contexts so it will not be included herein.

By 1979 I knew that the prevailing whites had no problems with "veggies" – for us. While the new system was not yet in productive play I really knew all was well for the future. Temporarily we could use remedial methods (discussed earlier). The red varieties were, however, an entirely different matter. I was sure that the new (yet still theoretical) system of growing would give substantial improvements but it was years and years away. Varietal compatibility analysis had to be worked upon. By this I mean varietal compatibility with the general climate and soil conditions of our new region. Initially, these tax-shelter vineyards had had their varietal plantings *selection* done by price-per ton by variety in the state annual report! I kid you not – that is how it was determined what to plant! There were some glances at Davis' Region Table but minimally.

We needed to explore each commercially viable variety, find those that were enough compatible with our climate/ soil situations such that we could fine tune to excellence with our winegrowing practices.

Thus, it was necessary to take each variety of interest and analyze it not in general but in detail. This process would take years and was intimidating to say the least but do it we must if we were to build a region of excellence. We also needed to look at non-mainstream varieties that had the *potential* for success in our area. Perhaps we could develop region fame eventually with one or two or three of these. But – which ones?

Books, books, books – maps and maps and maps ad nauseum. I read everything I could get my hands on from the French and Germans (nothing of importance in Spanish yet later one of the very best books around is from Spain – "Tratadura de Viticultura" by Professor Hidalgo) – focusing on their cooler areas. Hugh Johnson's "World Atlas of Wine" was fantastic – those maps, *not* the text. And, of course, tasting wines; comparing California efforts with those of Europe was a constant. I suspect that I am among that lucky group that has spit out more great grand crus than most people have drunk. But I was limited to the west coast California focus and our supply of fine European wines was very limited. (I do give praise to Kermit Lynch's findings, though). Thus, I made forays to New York City and Washington D.C. where the foreign wine selections are awesome. My true education began there aided by Peter Morrell of Morrell & Co., Mike Goldstein of Park Avenue Liquors (not on Park Avenue!) and Pat Cetta of Sparks Steak House. These men showed me wines and varieties I had never heard of. I think it deeply amused these city sophisticates to teach this "country bumpkin" who was so hungry for knowledge but really so naïve. Many nights Pat Cetta and I would sit at a table in his restaurant past closing while he taught me by pouring Italian wines. One night, with the staff semi-circle around us, I said to Pat that the kitchen is closed and we should go. He growled "The kitchen is closed when I say it's closed". We drank wines for several more hours and I never again brought up that subject! Pat also gave me a pearl one evening as I let slip one of the California mantras of the time about how well one of my wines would age. He said "Well, you know Doug, when my customers come in for a steak they want a wine to go with their steak! They don't want to take the wine home and age it". Wow – what an insight! That also led to another study area.

These experiences, and many other people's contributions (too many to mention), were priceless to my early fundamental comprehension of the European "wine" side of my learning and "sorting out". You must understand that these were people to whom every wine salesperson of the world was bringing samples. Their knowledge was incredible.

But, books, maps and New York City were not enough for this farm boy. Farm boys need to *see* the dirt, feel the air and smell the smells. Most importantly they need to talk with other farmers, already experienced, about what they do.

Only a small percentage of European wines make it to New York City. Most are consumed at home, particularly the lesser known but highly cherished ones. The small-lot wines definitely are local-only. Even today the garagiste wines of France and the specialty wines of Spain, Italy, Germany, Portugal, etc. don't make it here. The situation is the same in California, Oregon and Washington – everywhere. It was in Europe where my next (or parallel) stage of exploration had to occur. It was there that my exposure was broadened immensely.

As I looked at various varieties and considered them for our use I initially had one over-riding criteria (and still do!) – did I *like* the wine from this variety? I mean – *really* like it. Then, I explored what I thought would be its amenability to our situation. I also – further along in the analysis – evaluated what I thought *might* be its commercial viability in the American market – not present but future viability. Some vine names simply won't work for us. Various thought aspects will be brought up as I discuss specific varieties upon which we worked. The varieties I will discuss in depth will, at first glance, seem off-the-wall wrong varieties for a cold windy region. But, as they say – things are not always as they seem. That is why we have tools for analysis. Surprises often occur!

TEMPRANILLO

Tempranillo (TEM-PRA-NEE-YO) is the great red grape of Spain. I first cut my teeth on the wines of Rioja – the historically noted area for quality red wines – when they were made in the old Spanish style. That style involved long periods in barrel then in bottle resulting in wines that most moderns would not find particularly charming. I, however, took a liking to the finer ones. Tempranillo is a very long aging wine and is very stable. It develops a silky finish over time. Its flavors have been often compared to France's Pinot Noir and Italy's Sangiovese. The better wines of Tempranillo are more on the elegant side rather than the powerful. In the past the Tempranillos of Rioja were mostly blends. Today, more and more are made as 100% varietals. In Rioja there are three authorized blending grapes: Garnacha (a native Spanish grape called GRENACHE in France and the most commonly used); GRACIANO (a minor blender used for touches of aroma and flavor); MAZUELO (a very robust grape known in France as CARIGNAN).

Subsequently the Ribera del Duero region came to the fore in some quantities such that bottles reached me and I fell in love with the wines. The Ribera del Duero producers often blend in a small amount of Cabernet Sauvignon (a marriage made in heaven) to the exclusion of the others. Today, a few Rioja wineries produce a comparable blend and they, too, are superb but few and far between. I like both approaches to the grape – but, then again, I can't think of anything about Spain that I don't like. I have been fortunate to have existed during the time that Spain emerged from nearly medieval times to a full-fledged 21st century country. I observed the EU pumping Euro-dollars into Spain to upgrade its infrastructure and economy to EU standards. In many ways I miss old Spain but the people sure don't.

Ken Gingras (mentioned in the acknowledgements) is the guilty party for my becoming a Spano-phile. He is one also. We and our wives have spent long periods wandering Spain, Ken as a bull aficionado and I meeting winegrowers, making friends and learning from them.

Very early on it struck me that the Tempranillo is an exceptionally versatile grape giving lovely, though different, wines from a wide variety of climates and soils. Depending upon the local conditions the farming techniques ranged widely though some were just hold-overs from a thousand years (more) of poverty. The Tempranillo grape grows from the cool hills of the Pyrenees through the hot and arid plains of La Mancha (central Spain) to both the west and east. It made its way down the Ribera Del Duero into Portugal (there the river is the DOURO) and took on the name RORIZ – the name of the estate where it was first planted – supposedly. It has become one of the two most important varieties for high quality PORT wine (along with TOURIGA NATIONAL).

Over the years, I owe a debit of gratitude to so many Spaniards who willingly shared knowledge about vines and wines (not to mention taking us into their homes and culture and making us family). I will mention a few. Sr. Javier Ochoa of Ochoa winery and also the founding director of EVENA, the research institution for the State of Nevarra was profoundly helpful to my understanding of the Tempranillo's performance in relatively cooler situations. He showed me and explained the projects underway at EVENA and the reasons therefore. He patiently explained to me the methods they used to change the habits of centuries to improve the grapes and wines of their region. We had long discussions about clones of Tempranillo and their merits/demerits. Another is SR. MIGUEL MORENO of Briones within the Rioja region. Miguel makes wines in the old Rioja style that are delicious. He is also a worldly wine merchant representing many Spanish wineries to importers. He has a vast wine knowledge that overwhelms me at times. Again, we've spent much time in discussing (and Miguel opening wines to *show* me) vines and wines. Tom Perry is an ex-pat American, married to a Spanish lady, who is head of the Rioja export commission. Years ago I tasted Gundlach-Bundshu's Tempranillo and promptly bought out Jim's last twenty-five cases. I loved it and, to the best of his recollection, there was maybe 11% Cabernet in it. Five of those cases were taken back to Spain by Tom to show Rioja winegrowers and to encourage Cabernet usage. Of course, just down the road a few hours was the variety already in use but that would be like taking a Paso Robles Cab to Napa to show them what they should be doing. Nothing really came of that. I think those few in Rioja using some Cab did, in fact, look at Ribera del Douro's market appreciation. The folks at Barbadillo near Jeres de la Frontera (actually in SANLUCAR de BARRAMEDA) instructed me in depth about their wines and vines for general educational purposes – Tempranillo is not grown that far south.

Repeatedly, almost uniformly, I was warned by growers and academics against the "Davis clones" of Tempranillo. They were of the opinion that they were inferior. Keep in mind that this opinion was formed long ago and in a place far away. Also, Tempranillo was not a mainstream grape (nor is it yet today), and better winegrowers did not work with it and our own winegrowing was juvenile compared to today. So – there are a lot of explanations for these assertions. Yet, I think there is some merit to their opinion. Davis's original clones (still on the books) were from Valdepenas – right in the heart of perhaps the European continent's hottest area! Further, I don't think anyone has ever asserted that Valdepenas grows superior wine. The vine has grown there since time immemorial and has adapted to that area. Why would we expect wood – *randomly* selected a century ago to provide us with quality wine in cooler climates? We Americans just haven't spent any time and energy on clonal work on this non-mainstream (for us) grape. The Spaniards have – to them it is THE mainstream grape.

In the past Spain was revered as a cheap bulk wine supplier to France and Italy. And it was so. Very little high quality wine was produced and that which was was mostly consumed in Spain. The production of quality product bottle finished really only began about 25 years ago or so. Yes, there is Miguel Torres of Penedes. Now, name another in Spain for table wines of which you were aware twenty-five years ago. Twenty years ago? Fifteen years ago? Ten years ago? I'm sure you see what I mean. That *beginning* has now *exploded* onto the world. Often the wine called UNICO (from Vega-Sicilia in the Ribera del Duero) is listed at the top of the finest wines of the world for the year. The fine Tempranillos wines of Spain are now marketed throughout the world. It may have taken those of us early promoters of Syrah twenty years to get the word entrenched in the American language but I'm sure it won't take us that long for Tempranillo because the Spaniards are doing such a good job for us!

This is another grape that, like Syrah, was perceived by most as a warm climate grape. Like Syrah it is mostly grown in those warm to hot climates. Valdepenas is scorching in summer. Ribera del Duero is regularly over 100 degrees F. Rioja is cooler though not by a lot in Rioja Baja. Rioja Alta and Rioja Alovesa are significantly cooler as the cool air from the gaps in the Pyrenees makes its way from the north, sliding down the mountains. The grapes from those upper areas and those from slopes in Nevarra interested me and attracted my attention to the adaptability of this vine.

Here at home armed with my observations I began to contemplate Tempranillo's possible adaptation to Monterey's climate. By this time our new vineyard system was in place and at production. I knew that we could farm some warm climate grapes with it – witness Syrah. Cabernet and Merlot were coming along nicely. I knew Grenache was do-able here. I focused more on Tempranillo and embarked upon clonal work, the methods of which I will not go into as they are proprietary. I purchased cuttings from various sources including Jim Bundschu's old vines that are right in front of his winery. He had no idea of what "clone" as they were there when he was a child. There are other old plantings throughout the state.

Along the way, FPMS released a new clone called MADRID 2001AG – I believe in 1994, if memory serves. Some years back there was some confusion about this clone. The Spaniards were very angry about it. Apparently some company pitched the vine to them as the greatest thing since bottled beer, and sold lots of plants. As time passed the vines would produce "only leaves no fruit" or so I was told. The company was sued and forced into bankruptcy. When I mention this clone they still get angry and I'm not sure why. Perhaps my Castillian Spanish isn't good enough to gather all the currents. I did ask why they planted a lot without testing. The answer was simply a tight jaw. I had planted some of this clone on the Ventana. I was afraid I had made a horrible mistake from what this Spanish professor of viticulture and a winegrower had told me. After a few months of periodic thought I went back to Spain. I had some questions. I raised these points to the professor: It had occurred to me that I had only seen Tempranillo spur pruned any where in Spain. Some varieties and clones are unfruitful in their basal area buds. Had anyone tried leaving a fruiting cane – or "kicker cane"? The professor looked at the others with him and then said no – no one had tried that but that that wasn't the problem. He said that the problem was that the anthers were "bent" away from the ovules so pollen wouldn't reach them! Reggie Hammond (director of production at Ventana – both Vineyard and Winery) and I looked at each other. We'd never heard of such a thing. The professor said he had taken pictures of this phenomenon. Needless to say, a few months later at bloom in Ventana Reggie spent a lot of hours with a glass studying anthers. He found no such condition in our vines of this clone. This clone has been fruitful every year. In fact, we have to drop fruit it sets so well. We do use a fruit cane but that is because of vine spacing and for fruit distribution in the trellis. The spur buds are fruitful and set. Frankly, I don't think we are talking about the same clone. I haven't followed up on this and I don't know where it stands today. The Madrid clone is useful plant material.

Out of our other work there are two clones of stronger interest grown on the Ventana today. Some of this wood has made it to other vineyards in very small quantities for testing work. These clones are very cleverly named VT-1 and VT-2 which means Ventana Tempranillo number one and number two. We spend a lot of time coming up with these clever names! Some of these vines are of sufficient age to have generated fruit of merit warranting going to battle. The 2004 MEADOR ESTATE TEMPRANILLO is the first bottled wine from these clones. The wine is roughly half and half from each clone plus about 10% Cabernet Sauvignon. It is a stunning wine in my humble opinion. Yah-right! It is a demonstration wine – meaning I hope other growers learn the merits of Tempranillo from it – or, at least, it tickles their interest. Prior to 2004 any grapes were put on the ground early or used elsewhere. Under the Ventana label there is a 2005. There will be no more under MEADOR ESTATE.

This wine is the culmination of a more than twenty year odyssey.

In my opinion at this early stage I think VT-1 will be America's premium clone of Tempranillo for the next half century or so. It does not tend to over-crop itself severely though it still needs some fruit drop. While Tempranillo is a big cluster big berry variety compared to Cab the VT-1 is smaller berried than the Madrid or VT-2 (they are grown on common soils). The wine is a little more concentrated because of that skin to juice ratio. The flavours are marvelous and pronounced which is what we would expect from cool conditions.

I'm sure that many people in the future may opt for VT-2. It is a heavier producer – wanting to set a lot of crop. In our close vine planting this is a problem and requires two passes – green dropping and a post veraison pass for fine tuning. Perhaps more distance between the vines could reduce this somewhat by giving the vine more "drag" allowing it to carry more. Time and experimentation will tell. For me the VT-1 fits the design better and I *think* the wine is better. One has to be careful when evaluating the long term merits based on juvenile vine performance.

As far as "veggies" goes, normally, if a vine has them it is more pronounced in a young vine. We have seen no veggie-ness in the Tempranillo – *not even* veggie precursors that disappear as the grape ripens. That is one of the aspects I look for in European vineyards when considering a potential varietal for exploration. If I find it in any real degree I shy away.

The fermentation practices on this varietal are pretty straight forward. We used 6 ton open top vessels with hand punch downs several times a day. We pressed when dry using a couple of light squeezes so we had no "press wine" as such. The wine went to tank to complete M-L on the gross lees. When finished and settled it was racked to another tank, SO2 added in process and then barreled down to rest. No SO2 was added prior to completion of M-L at any time. The yeast was BM-45. Really it was just simple winemaking. Reggie Hammond and Miguel Martinez determined the percentage of Cabernet at the time of assemblage. Actually, they did all the work and art but, as owner, I get the credit. Funny how this world works.

I am of the opinion that the Tempranillo grape will become very important to our industry, perhaps not to the level of Cabernet or Syrah but important none-the-less. It gives a world class, beautiful wine as the Spanish have shown. We and others have shown that it transplants well to the west coast of the United States. Not only are we enamored with its performance on Ventana with these newer clones, the old clones are presenting lovely wines in modern hands. I particularly attract your attention to ABACELA winery in the Umpqua River Valley in Oregon. Their style is the old Rioja style using the classical blenders Graziano and Mazuela. It is delicious as a young wine but it will also gain with age. It provides a different "taste" thus a wider choice range to the consumer and it is a lovely food companion. It harvests at a little offset from Cab in the same location thus providing the winery better utilization of equipment and facilities. It is a cross blender with Cab, Merlot, Grenache and Syrah bringing something to the table with each thus providing the winemaker another tool and a wider potential range of wines for his product line.

We at Ventana worked with and actively promoted Syrah, Grenache Noir, Merlot and Cabernet Sauvignon. Tempranillo – as a cross-user with these varieties was such an obvious variety of interest to have *if* it were amenable to our area. It substantially increased the range of possible proprietary wines one could produce. Another aspect of this "range" of tools is that in different climate years different flavours may develop – some good, some not so good. The varying cool climate affects different varieties differently and part of those effects depends upon *when* said adverse weather occurs. Thus, in given year a lesser quality in one variety can be offset by a greater quality in another. The more useful varieties there are the greater the mathematical probability of producing a superior product *every* year or nearly so.

This is exactly what a *winemaker* (in this case an accurate term) is doing when he makes an assemblage bench sample for evaluation. This process can take skilled blenders hours, days or even weeks. One mustn't fixate on just one palate on one day. Proposed blends are built, poor ones discarded (paper trails of percentage blends are kept) better ones kept, and Reggie and Miguel taste and evaluate. The process is repeated several times days apart. Why? Our personal body chemistry varies as does our sense of smell and taste. We all have off days thus the repetitions. In the past I would suggest varieties and percentage ranges of each as a starting point and Miguel or Reggie would play with those. After the whittling down they would present choices to me that had made their "cut". There may or may not have been agreement on first choices by each but they never presented me a "dog" – okay, maybe one or two to show me what a lousy suggestion one of my original recommendations had been. That was okay – that, too, was a good teaching tool and they had the full freedom of action to so put me in my place. If there are restrictions on freedom of expression there is limitation on learning and progress. Some winemakers have sufficiently overpowering egos to not allow this approach but we find it very useful to consistently generate the very best wine of a vintage that our *team* (vineyard and winery) can and avoid the mistakes or off days by a single individual. We also share the praise – or blame.

For a few years now I have not been part of that final step (except in this one Tempranillo because it was so dear to my heart). Both Reggie and Miguel are far beyond me now when it comes to red wines and they know even better than me what will give the artistic style I desire.

SYRAH

This adventure began in 1974. Walter Schug – a young man from Germany – had been Gallo's buyer of grapes for some years and had become familiar with grapes growers fondly called their "Pets". In 1974, Walter was the winemaker at Joseph Phelps. Walter rather liked the grapes from a tiny portion of one of the Christian Brothers Napa vineyards. Brother Timothy was kind enough (he was an extremely kind man) to part with some of those grapes he called "Syrah". Thus, the first labeled SYRAH was produced from that vintage by Joseph Phelps Winery. That block was, in fact, an assortment of grapes as was common in those days[3]. Those grapes later were torn out as they were badly virused as were the ones planted at Phelps taken therefrom. Sadly, those vines are lost to us – we have no idea of their genetic nature.

That same year the major player in the modern pure Syrah drama entered the stage. Gary Eberle of Paso Robles fame (and offensive tackle of Penn State fame under Papa Joe) *earlier was* a member of what came to be called "The great class "of graduate winemakers from U.C. Davis. Among them was an Australian – one Paul Crozer. At their informal gatherings said Australian kept bringing bottles of some southern wine he called "Shiraz" (Syrah in Australian).

Gary became enamored of the wine eventually and studied the variety intensively. Later (1973), in the real world, Gary was tasked with pioneering, to a large degree, the Paso Robles area with the large "Estrella Vineyards" project. Given the moderately warm Paso climate, Gary was adamant that some Syrah be planted. He prevailed – it was. In conjunction with, and help of, the late Curt Alley, Gary was able to "acquire" some wood cuttings from a *non-released* test vine (from Austin Goheen's "Vit block" existing only for varietal identification comparisons. The vine was virused)

---

3 Verbal to me from Walter Schug

referred to by Professor Alley as the "CHAPOUTIER" clone of Syrah.[4] At that time Chapoutier had only the one vineyard above TAIN in the Rhone Valley so we assumed it came from there. Further research by me disclosed that Goheen got the plant from Montpellier University and he referred to it as the "Montpellier" clone. However, Montpellier did, in fact, get that particular material from Chapoutier!

In early 1974 Gary gave those pieces of dormant wood to me to expand it out by a process called "Mist Propagation". In this process we plant the dormant wood in pots, grow them out in greenhouses, cut the green "runners" or canes into one-leaf snippets, root them under misters, transplant to little pots, grow them to size and send them to the vineyard for field planting – all for a fee, of course. Thus, in 1975 Estrella was the first modern American planting of the Syrah grapevine. Gary Eberle was the single driving force behind this planting and the visionary who recognized the variety's potential.

To the extent Americans thought about Syrah – which was nil except for Schug and Eberle – it was construed as a warm-climate grape variety. This was because the Rhone Valley, Cotes du Rhone, Chateauneuf du Pape, the southern areas to, and into, Spain and even in the area of Toulouse were its territory and those areas were viewed as hot.

Eberle's fixation attracted my attention. He is quietly a very intelligent person and when he does speak he should be listened to carefully (typical offensive tackle – quiet but effective). In Monterey we are *much* colder than Paso Robles and we have a driving cold wind nearly every day directly off Monterey Bay (Pacific Ocean – Japanese current fresh from Alaska). The winds were shredding the leaves of some varieties – particularly a block of California Petite Syrah I had. This is a different variety not related to Syrah at all – different parentage. DNA analysis has finally shown us that in about 2000. I thought about Syrah off and on. I had put a few vines in the ground in 1975 – culls and excess plants

4 All verbal to me at the time and again later by Gary Eberle

from Gary's order but paid them little attention – they still weren't up on my radar scope of thought. I don't know today whatever happened to them. I think I removed them for purity reasons. One day it occurred to me that Syrah's finest performance historically (at least by reputation – even as a blender into Cabernet in Bordeaux!) was in the Rhone Valley itself and the Rhone has a very famous, sometimes violent, wind called the "Mistral". This wind is so violent at times that houses in its regular paths have no doors or windows on the north side! Thought: this grape variety had evolved in and adapted to, this wind. Perhaps it was "wind tolerant".

The second, subsequent thought was the source and nature of the Mistral. In the south the very warm wind coming off the Mediterranean is called the "Marin". The really hot one coming up from Africa is called the "Sirrocco". The Mistral, on the other hand, blew down the Rhone Valley. It begins in the Alps. It is a cool, sometimes cold, wind! Voila or Eureka or whatever! From this epiphany of thought came the conclusion that Syrah MIGHT – just MIGHT – be amenable to Monterey's "Mistral" which was also hard and cold. Testing had to be done it was clear.

In late 1977 I contacted the Foundation Plant Material Service (FPMS) – our repository of varietals at the University of California, Davis – and inquired about Syrah. The FPMS had recently released a known-virus free clone of Syrah acquired from Australia and called it SYRAH/SHIRAZ. By this time Professor Austin Goheen would allow no release until a plant tested clean – and this was Goheen's empire, he controlled it. In the interest of the widest and fastest dissemination of clean material FPMS had a policy of giving first call on material to nurseries. Growers could get FPMS material but only that left, if any, after nursery requests were met. I had to wait until "nursery season" was over. Later, when I called again I asked for the "twigs" and "tips" – the leftovers. The lady – whom I knew from all the nursery work – asked me "Doug, why do you want the twigs?" I said "Because the nurseries have taken the wood". She said "No one has ordered any. You can have all the wood". Yahoo – God *does* look out for farmers, fighter pilots and fools and I touch every base. We still have the "purple" tags that came with certified mother plants directly.

So – it was in early 1978 that I had enough wood to graft over certified Cabernet vines of a little over 1.5 areas of old-style vineyard (518 vines/acre at 12' by 7'). These vines were immediately adjacent to my block of California Petite Sirah – deliberately placed there for comparative purposes.

The Syrah vines absolutely LOVED it here. Vigorous, wind-tolerant, ripe fruit, beautiful flavours, easy to farm – a winegrowers dream! The California Petite Sirah? Shredded leaves, prone to mildew and Botrytis, difficult flavours (remember – this was old style format. Perhaps it might have been different new style but I seriously doubt it), hard to ripen, etc., etc. Not a farmers dream – well, maybe one kind of dream!

Oh – it wasn't all pure joy. There is a learning curve on how to farm it here and there are some difficulties but all within normal farming capabilities. Handling this variety has pretty much stabilized now and I understand moves to make in given situations. Depending upon the weather tipping may need to be done earlier or delayed, thinning may need to be done once or twice depending upon set and leaf ratios.

In France, there have been significant instances of difficulties with rootstocks and growers intending to plant Syrah on rootstocks should thoroughly study this aspect. I cannot give practical guidance because the plants on Ventana are own-rooted except that first 1.5 acre portion which is on Cabernet roots – still vinifera. That block was quickly interplanted after grafting to six foot between row spacing and in those rows the vines are 4.5 feet. Some interplanting of the early rows has also occurred – all own-rooted.

There have been problems here in America as well. I suspect the "ooze" problem and/or vine declines reported in third or fourth year are phenomenon related to the problems in the French literature. It is being "studied" here and it is being looked into in France. I'm not sure there has been collaboration yet and I think the answers are still a mystery to be unraveled. About the only comment I would make is that a grower should look to known and locally experienced clone/rootstock compatibility *before* all the normal rootstock considerations.

At Ventana we made ourselves (our normal practice before selling any grapes) many "test" wines commencing in the early eighties. These were lovely rascals though mainly for "learning" and for "show and tell". Subsequently, River Run Winery worked with me over years making many gold medal wines from the grapes beginning, I think, in, 1985. By 1981 or '82 I was publicly asserting that I thought that Syrah would be Monterey's first truly great red grape. Of course, once again I was informed that I was obviously in need of serious psychiatric care – that I was nuts – that *everybody knows* Monterey can't do reds! By 1988 I was so enamored of the grape that I cast all caution to the wind and said "no guts, no glory" and decided to increase my production 100% - I planted another 1.5 acres! Heroism is hell. Later we planted considerable more acreage.

It wasn't until 1989 that we saw the next planting in Monterey County by another grower.[5] Nowadays Syrah is planted throughout Monterey and doing beautifully. We have supplied the wood for much of this expansion. Many of our local wineries produce Syrah wines and awards for this grape roll in. The consumers are loving it.

The exploratory development period is over and the expense incurred. The Syrah is now widely planted throughout the state and is thriving. Establishment of the word in the public's mind as a wine of merit probably took longer than most of the discovery mode! It is now entrenched in our lifestyle and taste patterns.

---

5 Verbal to me by Rich Smith on year he planted his "couple of handfuls of vines" – which I assume was "ten".

The Syrah follows a pretty straight forward winemaking pattern in our hands giving no particular problems. We destem into 6 ton open top fermentors adding no SO2. We skin soak about 12 to 18 hours and add yeast in one corner of the fermentor. We punch down several times a day and lightly press with a couple of gently squeezes to tank where we finish M-L and settle out. The M-L bacteria were added at about one-half sugar fermentation. At the completion of M-L SO2 is added to 33 – 35 ppm during the racking to another tank then the wine is transferred to barrel.

Syrah has a liking for American oak if the barrel has been properly constructed – meaning air cured not kiln dried. We prefer all Minnesota oak even though more expensive. Depending upon the year the American Oak ratio to French will vary and within the limits of necessity but we like Syrah with more American than French – an artistic decision.

In the thirty years we have grown Syrah we have learned a huge amount about this grape. It still has some tricks to show I suspect. But also as these old vines are now presenting us with essentially a different grape than when young we are having to learn to deal with that and re-focus our art and work towards that new expression. The winegrower must have knowledge of the end goal – i.e., the intended wine style. Then, the vine must be "broken to harness" or trained and controlled to that end. At twenty years of vine age those grapes went into a wine labeled Meador Estate Syrah "Maverick". That 1998 Syrah not only was the recipient of a myriad of gold medals here but also was awarded gold in Paris with French judges. Not too bad for a county that "can't" do reds.

This is an easy variety on which to grow "tonnage" of red wines. It is more difficult to grow "Syrah" wine of superior merit. There can be large annual variations – particularly with youthful vines – that must be controlled and balanced to the prevailing weather patterns. There is no substitute for long experience with the variety on a given site.

GRENACHE NOIR

I love this grape so historically maligned and misunderstood by Americans. Happily we are slowly acquiring some small appreciation that will grow significantly over the next couple of decades – it is becoming important rapidly now.

This grape is a native of Spain where its name is GARNACHA and it is grown throughout the country except the far north and the far south. It is the main varietal for the lovely dry rosados so delicious in that warm clime and such a great companion for their foods. It is used in fine wines as a blender and as the primary variety in wines with its name. It produces big bodied, dark-colored and, relatively speaking in European terms, generous or higher alcohol, and aging ability when assisted by other varieties. It has a strong tendency to oxidation when not accompanied by another variety to counteract this feature.

Long ago the vine moved north into France and the French, not liking anything so harsh sounding, gave it the name Grenache which it carries there and here today.

The wood originally brought here was just collected in a vineyard somewhere where has been lost in history. The Grenache is a generous yielder and was favored by the old folks. With the wood brought here it turned out that it was mostly only good for red but again that was partly a function of our culture. In the mid-1900's it was ubiquitous, appearing everywhere in the gallon jug format under the name Grenache Rose. It was deep, very sweet and tasted good – almost as good as Coca-Cola. As America moved into its 1970's and 80's fascination with food and wine fads that came and went with their associated snobism and one-upmanships, Grenache Rose was one of its victims – probably rightfully so as it was. However, sometimes we tend to throw the baby out with the bathwater and that, also, was the case here. Combined with Cold Duck, Mateus, Lancers and others of the time, in California it became tres gauche to be seen with anything pink in your glass. Eventually the disdain also applied to anything with even a trace of sweetness to it. Grenache became completely ignored as a varietal of any potential merit and, essentially, abandoned to the scrap heap of history. Oh – our ignorance!

To a few of us, however, Grenache was still on our radar scopes but down low and very faint. As I mentioned under Pinot Noir discussions I was very fortunate to be taken under-wing by the early Pinot warlocks – David Bruce in particular. At one gathering at his home, while Pinot and Chardonnay were the focus, David opened a bottle of his 1969 Grenache just as a palate break. I was stunned. The wine was black, thick and utterly delicious. That wine really began my serious trek into the Grenache world. A funny little story about that wine occurred a whole bunch of years later at the charity event near Paso for the local Public Broadcasting of Archie McClarens. One part of the event was held under a huge tent with dinner tables scattered under it each hosted by a winery. A table off to the side held additional wines of each winery as replenishments were needed at the table. As I was returning to my table on one foray Randall Graham (Randall was just coming into play with his erudite marketing – a very new kid on the block) stopped me and wanted me to taste a wine of his. I did so and said "Randall, you rascal, you've slipped about 7% Grenache into this little beauty". He was stunned – accidentally I'd hit the right percentage. He said – "How the hell can you tell that?" Then I played with him – I said "I'll tell you something more – you got the grapes in Gilroy, probably that old vineyard on the north of the highway to Madonna!" His chin was on his chest. Then I winked at him, smiled and continued back to my guests who were dying of thirst. Shortly, Randall appeared at my table demanding to know how I could do this – had I been to his vineyard or talked with his people? I was done playing with him so I said "Randall, in 1969 David Bruce made a Grenache from the vineyard I described. If one ever had a taste of that wine one could never forget the aroma. Randall then laughed and said he, too, had tasted that wine and that's what led him into playing with Grenache – and, yes, the grapes had come from that vineyard. There is a point to this story and I'll come back to it later when I discuss clones.

My next experience with Grenache occurred in Monterey County. As I mentioned elsewhere, the original choices of varietals by the original promoters was a bit off-the-wall. The original Monterey Vineyards vineyard complexes were mostly around Gonzales. On upper Camphora road (to the east of Gonzales) towards the Swiss Rifle Club there was a small block of Grenache as well as various other strange varieties. Roy Thomas of the old Monterey Peninsula Winery (now long defunct) was very much into innovative use of these varieties but at *very* high sugars. December would come, leaves gone off the vines, grapes hanging and there would be Roy gathering some grapes. He made some unique wines and had a loyal local following. After Roy finished with his tiny lots I would observe these lonely grapes still hanging. Somewhere in late December a machine would lumber in and harvest the grapes for Monterey Vineyard Winery – sort of a "well, we've got a little time, let's bring them in" sort of thing. There was no market for them. These shriveled down, wrinkly, not pretty, essentially abandoned things were not really on anyone's interest list. They would be fermented in harvest gondolas (covered in plastic) outdoors. This went on year after year. The funny thing is – the wine was lovely! In those years when reds were such a problem the few red wines that were drinkable had that Grenache blended in them! Eventually that block was torn out. No one really made the connection except me (I was somewhat privy to their operational knowledge). Thus, I knew that Grenache had the potential of superb *flavours* in Monterey but the color and depth factors were still a problem. I had other, more important, things on my plate and for quite awhile that knowledge was sort of in hibernation – there but not pushing on me.

Somewhere in the early eighties I became enamored of Cote du Rhone wines particularly GIGONDAS as an outgrowth of my studies of Pinot Noir. It had bothered me why negociants in Burgundy had Grenache in their portfolios when Pinot Noir was the only red variety authorized in Burgundy. Further, then when my palate was much more accurate than now, I often thought I found *favorable* aspects in some red burgundies that were Grenache in tone – particularly in lesser years. I began buying some of those wines and studying the area as well as Chateauneuf du Papes. I'm not sure that this old farmer's suspicions were misplaced and that all I was sensing was unique "terroir". However, in the early 90's, I was at a winery in Gigondas where a tanker truck was being loaded with bulk wine. I wandered over to the driver and asked where he was going. Beaune was his answer. Some years later I read in Robert Parker's book on the Rhone that he was raising the same questions. I'm sure we are both wrong in our suspicions! Right.

But – back to Grenache. The disdain for pink caused Americans to forego one of life's pleasures. We failed to note that nearly one-third of the table wine production in France, Spain, Portugal and Italy was pink – Rose (France), Rosado (Spain) and Rosato (Italy) and was/is a very important part of their culture. On a summer day one finds oneself wandering into a small village tired and a bit sweaty and sees a café with one open table left under an umbrella and grabs it. After about a nano-second of contemplation one then orders a cold bottle of rose, some bell peppers sautéed in olive oil and garlic and some bread. Is that one possible picture of Heaven? It is to me. That scenario occurs all over the place. Well, why the popularity of pink wines there and not here? It is because theirs are bone dry! They refresh! They are low in alcohol! They taste good! And – they work with a wide range of the foods of each culture. Happily, in very recent years Americans are starting to find this out. And they are enjoying.

Back in 1990 I planted some California Grenache on the Ventana. That planting was not for the serious Grenache that came later. This was specifically for Rosado – a dry rose. Friends had been grumbling after the bull runs in Pamplona that it was another year until we could get more Rosado. I decided enough was enough – I would grow some grapes and make us some Rosado so we could have it throughout the year. When we also started showing the wine in the tasting room it put severe pressure upon our original intent – a supply for myself and friends!

At the same general time I began to get very serious about Grenache and hit the books and discussed thoroughly with growers, both French and Spanish, the nature of the beast. It turns out that Grenache is like Pinot in that there are various colors! Like Pinot Blanc in the Pinot family there is Grenache Blanc in the Grenache family – it is the foundation grape for white Chateauneuf du Papes. As there is Pinot Noir (Noir = Black) there is Grenache Noir which is the main grape used for Chateauneuf du Papes, Cotes du Rhone, Gigondas, Vaqueras, etc, etc. and, as Garnacha, many of the Spanish reds. And, again, as there is Pinot Gris (Gris = Gray but really just a Pinot with very light color) there are Grenache clones with light and unstable color that are used for rose-type wines. It turns out that our ancestors simply grabbed – in general – the light colored wood material and that's what we've expanded here – for rose. The marketplace probably encouraged that as well. These small pockets here and there of old vines – virused, no doubt, that give wine like the Bruce '69 – indicate that some of the old-timers did bring in Noir strains and they were probably very prevalent in those old field-blend plantings. Of course, in those days nobody paid any attention to clones just varietals and even that wasn't of terrible importance. That good old field blend was the important thing. When I came into the industry in 1972 it was still a common thing to go to the Italian wineries in the Hecker Pass/Gilroy area with your own gallon jugs to be filled from spigots in the big wooden tanks. People had their favorites which probably were a function of what composed the field blend but nobody thought about that – they just liked this winery over that. Those old Grenache Noir clones have essentially been lost to us. If one of you young readers wishes to do something beneficial I recommend to you Pierre Galet's book and learn how to recognize the Grenache leaf then go around the State to all the old vineyards you can find and see if there are Grenache. If and when you find vines you might flag them and study the fruit. You might then go to the trouble to plant some even contracting to have them cleaned up from virus (if you are rich and very young). Some of those old clones could be diamonds in the rough. At least that 1969 Bruce would so indicate!

The Europeans, however, did not lose their interest in Grenache. There exists a wide range of clones of this varietal well studied. We (Americans) have now legally imported many of these clones and they are available from nurseries. Many California growers have various clones planted and producing and there are more and more wines from them appearing on the shelves every day. Most, so far, are from warmer climates.

From my early observations I knew that Grenache would produce superb flavours in cool climate and I thought that the new vineyard design would aid in any ripening problems that might occur. One characteristic of Grenache that bothered me was its nasty tendency to "shatter" if the weather was not to its liking at bloom. It will also do this if too much vigor is present (excess nitrogen being one culprit). There were, I thought, farming techniques to apply to the vigor aspect but what about our kinky, unpredictable spring weather? Finally, I decided the hell with it – we'd just accept a crop failure from time to time and not plant so much that crop failure or shortages would hurt us. We'd handle the wine such that we could "bridge" a year if we had to or just run out. We planted. The Ventana has the 3 clones that ENTAV-INRA (the French clone controlling entity) ranks as number one for wine quality and another clone (#220) that

I call an "insurance" clone. The French, being French, have these lesser wine clones that have, supposedly, lesser shatter tendencies. That way if your really good stuff shatters one still has a crop and "we'll call it a lesser year". Funny thing – our observations so far are that the superior wine clones have less of a shatter problem then the one that is supposed to "protect" us. Mother certainly works in mysterious ways. It may be just the roll of the dice, though. They bloom at slightly different times so weather on those different days may affect this. Our "differences" may have been due to our pruning practices. I would put just a few of my best pruners on the Grenache so the time of pruning each was slightly different.

Somewhere in the early 90's I rented a villa among the vines in the Cote du Rhone – outside a bump in the road called Mondragon near Bolene. Various friends, family and other California winemakers dropped by and stayed a while (the place slept 11!) among them Tom Burgess of Napa fame. I was fortunate that a very plugged-in friend – Jean Louis Tourtin – lived in Bolene and he was a fantastic "Pocahontas" guiding us to whoever and whatever I wanted to see. He opened many doors that may not have been so easily penetrated and elicited answers that might not have been so open. Toward the end one night at the villa he asked me if I had seen everything I needed/wanted to see. I said yes except that I had really wished I could have met Paul Avril (Avril is like the Robert Mondavi of Chateauneuf du Papes – everyone wants his time. He owns Clos du Papes) but I had received no response to my letter. Jean-Louis said "You've got to be kidding – you want to see Paul Avril?" Oui – said I. He grabbed the phone, dialed a number, rattled off machinegun French that I couldn't follow, held the phone to his chest and asked me "Nine o'clock tomorrow morning okay?" Oui – said I. He got off the phone and said "Paul Avril is one of my best friends". Wow – and so M. Avril was our guide through his vineyards, introduced us to his son, Vincent, guided us through the winery and not only was he our tasting host but also was the cashier for our purchases! He was very open about his winemaking techniques as were the people, I might add, at Chateau Beaucastel (their winemaking practices are different).

Again, the Grenache is a varietal that has adapted to the Mistral winds as it can, at times, harass the area. At Ventana it has adapted and performed well. We have seen severe shatter at times in the California Grenache planted for Rosado and we have seen partial shatter in the 220 clone but none so far in the elite clones.

The Grenache Noir really wants to overcrop itself and greed *must* be fought constantly. It is critical that in cold climate one plans for and follows through with at least two passes of fruit thinning! It is a big clustered variety with potentially large berries. If one allows a luxury condition it will turn those berries into plums – even bigger. That is not very good for the wine. It should be grown with a winemaker's attitude not a grower's seeking tonnage. The first thinning pass must be done soon after set to reduce the drag on the vine such that there is enough energy available for proper shoot extensions and foliage development. Failure to do this will become very noticeable when those seed generated hormones slam into the vine. It will come to a screeching halt! On the other hand one must not take off so much the vine compensates for excess vigor by growing too much cane and making the berries huge. One must have drag on the vine to keep these berries modest. Then – *after* veraison the second pass should be done to adjust final tonnage. The full chess board of the year is apparent to one, the size of the berries is controlled and all is well. On this pass ask the worker to eliminate clusters that are still green fully or partially because of late bloom or development. The color difference is apparent. Now – with this "KNIFE-INDUCED UNIFORMITY - the block is amenable to machine picking, if desired, or by straight block hand picking, if desired.

Grenache is a very straight forward fermentor in our experience. It is a precocious wine with beautiful aromas –

one in particular often occurring that I call "candy apple" which I like. It is a superb blender. We use it with Syrah. We use Syrah with it in making our Rhone style red wine blends. It can be blended with Tempranillo and it also can work well with the Bordeaux complex! I do not recommend this varietal as a stand-alone varietal. It has a proclivity for oxidation on its own and it needs at least small amounts of another variety to compensate. We have made some lovely renditions of this varietal under the "Beaugravier" proprietary name (Ventana and "The Boss" under Meador Estate). A few years ago according to California Grapevine's end of year summary of awards, the Beaugravier was the number 11 highest award-winning blended red wine for America – in a field of 468 blended red wines receiving awards that year. Ventana's "Due Amici" was number one!

This variety is highly recommended by me for cooler areas as well as its historically perceived affinity for warmth. It is in cooler areas that it will really "show off" its merit particularly in rocky soils. The French have a saying – "Plant Grenache in rocks". In the case of Ventana that holds true and that is our experience. However, I do point out that while Chateauneuf du Papes support that, there are many locations in Chateauneuf that are *not* rocky and the Cotes du Rhone is really not a rock patch. In Spain the variety performs well in both rocky and non-rocky situations – just differently.

MISCELLANEOUS

(A) PORTUGUESE VARIETIES

In 1990, en-route to Pamplona, Spain to play tag with some bulls, meet with friends and study some vineyards, I decided to stop along the way in Lisbon. I had never been to Portugal but to some peoples' minds I was somewhat competent on vintage ports. I certainly had an affection for them. The previous year at Stuttgart, Germany's equipment show I had met Peter Bright – an Australian ex-pat then managing the largest winery in the Setubal peninsula, an area just south of Lisbon. In the course of Peter showing LuAnn and I around one evening we met up at a restaurant in Lisbon. Peter brought two bottles of red wines – unlabeled. I was shocked that such gorgeous wines could be done in the Setubal – hot country – the home of Lancers and Mateus. Peter said "No, No – these are from the Douro". I had never had, nor had I ever heard of, dry table wine from the Douro. Peter Bright is now one of Portugal's rock star winemakers and he focuses on northern fruit. Jerry Luper of California fame is also shining in Portugal.

Contemplating those wines led me to decide that we needed to explore the "Big Five" (plus a few others) of their 40 some red varieties grown in the Douro. I was able to acquire four of those five – the one exception was Barroca. However, the exploratory purpose is not to grow "port" in cool climate but rather investigate their usefulness in dry red table wine production. They are probably useful as blenders, bringing unique, indescribable nuances to a wine. I encourage some young winegrower to pursue this area of adventure and teach us yea or nay. The earlier vines planted by me have been removed since the sale and the project unfinished.

(B) Chenin Blanc

This is another area I encourage one of the young to jump into. The existing clones of Chenin Blanc in California have been here a very long time and selected over time by apparent disease-freeness and yield-only! Chenin Blanc has become the jug wine grape of the central valley and, thus, déclassé just as Grenache was.

For a long time the last three producers of barrel fermented serious quality Chenin in California were Chalone, Ventana and Chappellet. There is now only Chappellet. These can be gorgeous wines but there is potential for far more superior wines from a few clones in the Loire Valley of France. I am no longer in the game but someone should seriously pursue this. One probably should use a proprietary fanciful name when the bottling time comes to avoid the 5 gallon boxes for 99 cents on the shelves that say "Chenin Blanc". But there are some incredible clones out there for some incredible wines. There was one in particular I bumped into years ago that I think was clone number 827 but I *very* easily could be remembering wrong. There was a time when I was hot on the trail of it and a couple of others, needing more tasting time there. I ran into a little impediment about then – I think they called it cash flow or something like that. What nonsense, but it did push the Chenin work to the back burner. Actually, it pushed it off the stove. *That* project is waiting for one of you young Turks!

(C) Others

There were/are a myriad of other projects of lesser impact or importance and not particularly germane to the new viticulture of Monterey. I will end this adventure in projects by relating to you a sign I once saw:

A good friend will bail you out of jail. A true friend is sitting next to you saying "Damn, that was fun". And I tell you – fun it has been!

Wine Judgings

God bless the wine judgings of yore! They were of immense value to us in establishing Monterey as a fine region – as they have been (and are) for many other emerging regions. During the period of pundits damning anything from Monterey with the broad tar-brush of "veggies" the blind-judging of wines was telling – putting fact to fiction and allowing recognition.

In those early days there were two commercial judgings of importance – The Los Angeles County Fair (LACF) the grand daddy of all controlled by Nate Chroman (wine writer for the Los Angeles Times) and the upstart Orange County Fair (OCF) ramrodded by Jerry Mead and the Orange County Wine Society. They couldn't have been more different! That was it. The California state fair system was for amateurs only – home winemakers, as I've mentioned earlier. Score a touchdown at either of these and wine shops and restaurants had to have the wine. At both the wines were served to *panels* of judges "blind". By "blind" we mean that the judges received the wines already poured behind the scene in coded glasses – the meaning of said codes unknown to the judges. The judges knew not the labels nor regions – only the asserted varietal. In general, varietal naming of a wine was an American thing – not European. Frank Schoonmaker introduced it here feeling that chateau–naming was too complicated for Americans! Chateau-naming and regional naming required knowledge not disclosed by the label. It was simply assumed one knew – if one cared.

Federal law at the time required that if a wine was named by variety then at least 51% of the grapes of the named varietal must be documentally in the bottle. It was not until somewhere in the 1980's that the present rule of 75% was installed. Even that rule was a compromise. A tiny activist group (5 people) of stockholders in Mt. Eden pushed for a 100% requirement (Pinot Noir lovers – Pinot and Chardonnay are normally 100% type wines). They were also from Chicago if I remember correctly – at least, the prime mover was. The argument in support that was made was that winemakers were "cheating" the consumer (for real) by putting something called "Merlot" (pronounced MER-LOT) into their Cabernet Sauvignon – thus degrading same! Merlot was cheap (what there was of it) and Cab was expensive! These thieving winemakers might even put some other un-pronounceable things in the wine as well! Of course, all this ignorance of winemaking bothered them not at all and they charged forward. Bureaucrats, with their love of hard rules, gave credence to the activists and released the idea of a new confining rule. The industry fought back bringing rational arguments to an emotional, vested-interest subject and at last settling for the compromise of 75%. Today we still see the effects of this misguidance (not necessarily bad) by the growth of proprietary fanciful names for blends – a throwback to the old chateau/regional naming. One just has to "know" or the back label may tell one. There are a plethora of new rules about front label naming or varietal identifications. Artistic freedom of expressing the best possible wine the winemaker envisions is again possible. The 75% rule just doesn't allow it in many instances.

At the judgings a person known as the "Chief Judge" organizes everything including the lineup of judges. That "lineup of judges" is a *very* interesting factor and *the* area of differences between wine competitions. The chief judge has tremendous power over the tone and outcomes of the respective competitions by his/her initial selection of "judges" and their panel assignments.

As the "grand-daddy", the LACF came along in a time of limited general knowledge of wines. Los Angeles was just emerging as a food center – still trailing *far* behind San Francisco in its worldliness, such as it was. LA was looked upon by Northern California as a barbaric alien nation far removed from any cultural exposure. It was Hollywood, nouveau riche, wannabes in all its *variant* forms. My god, when they came to San Fran some didn't even wear ties!

How gauche, open collars in "The City"! This was a time when one "dressed" to go to "The City".

Los Angeles was striving hard to move into the new food awareness world and, along with it, the fine wine world. Jean Bertronou probably started the serious theme with La Chaumiere in 1965 and then L'Ermitage in about 1975 attempting to rescue L.A. from its "gastronomic hick-town status". Wolfgang Puck arrived in L.A. in 1975 and it took a short time for him to dazzle.[6] Michael McCarty showed up from the east in 1979. Patrick Terrail opened Ma Maison in 1973 – and received terrible reviews for two years until he turned the kitchen over to Puck. In 1982 Puck opened Spagos. In 1983 he opened Chinois.

As you can see by these dates the L.A. rise to food and wine awareness is a relatively recent phenomenon initially occurring in the late 1970's and spending its adolescence in the 1980's. I lived through this period of L.A.'s emergence.

Nate Chroman was a powerful promoter of this growth both in his role with the L.A. Times and as a chief judge of the LACF. If Nate was interested in you or your wines or the wines had shown well at the LACF (and if you were properly deferential) you were given the "opportunity" to take him and his wife to dinner at Ma Maison and watch Terrail fall all over him. And it was understandable – his pen could do wonders. Not everyone could get that "opportunity" – certainly not repeatedly. Nate periodically gave a few of my wines "okay" write-ups but I think he was more fascinated by my heretical theories and discussing the foundations upon which they were based. It was certainly an "experience". Little did he know that each of those "opportunities" ate up most of my budget for the succeeding month! Nate was adamant that the LACF was for California wines ONLY. Many years later he and I had many phone conversations that resulted in the LACF being opened to all. I would strongly argue with him that we Californians could not fight for our place on the world stage if our most prestigious judging (some thought)

6 The United States of Arugula: David Kamp

would not allow us to go nose to nose with the world's wines! By that time *our* skills were such that we should have that venue. After all, the Judgment of Paris had already occurred and, more than a decade earlier though I think of marginal technical importance, we should follow by opening our competitions to all. I wanted to go into the arena and face the Tiger with California judges. After all, my market was California.

Nate staffed the lineup of judges for the LACF with an across the spectrum of wine – knowledgeable people he could get at the time. Some were cronies – but knowledgeable cronies – of his from the restaurant scene. Others were wine merchants, university professors, consumers with whom he was acquainted and a few – very few – professional winemakers and winery owners. This is an important distinction – this across-the-spectrum of consumers of FINISHED wines presented for consumption. With the exception of the winemakers (more on this under OCF) all his judges were experienced with wines ready for consumption or nearly so. Wine sales people presented bottled samples to restaurateurs and merchants not for aging but for selling. Consumers bought wines for consumption. Oh – many *talked* about aging wine but few did so in fact beyond perhaps a year or two. Long-aged wines were conversation pieces or bragging items – not for consumption. Often I have seen someone bring out a very old bottle of wine for prestige purposes, pour (often decanted hours ahead in the belief in that erroneous ritual), and then everyone would sing the praises to an absolutely unpleasant wine!

This format was very useful to the public – exactly what its purpose was. Judge blind wines tendered to the public and give the consumer some sort of independent quality evaluation as a guide for their purchases. Yes – that guidance was a function of the group prejudices about that very subjective matter of taste. Yes – those wine-knowledgeable judges were not representative of the general population's taste level. Yes – there were strong prejudices against certain wines, grape varieties and styles. Yes – often too many wines were served to a panel, palates became fatigued and awards given by default so they could go home (this plagues almost all judgings – often later tasted wines receive awards). Yes – there are all sorts of flaws varying according to particular judgings.

But – a big but – it is also a format that allows unknowns to rise from obscurity or allows famed wines to restate their superiority. It allows wines and regions to emerge unfettered by prejudice against said label or region. It allows different artistry and/or technology to be evaluated for merit within sort-of mainstream palates. If "newness" is too far avant-garde it allows progressively "training" the judges by exposure for several years until they "learn".

In contrast, the OCF is formatted differently. Jerry Mead (founder of WINO, syndicated wine writer and reviewer and is, now, sadly – a dear friend deceased), striving to give the new-kid-on-the-block instant credibility in competitions with the LACF, structured his judging panel as all or nearly all winemakers and/or winery owners and promoted heavily that fact. The subliminal message was that no one knows wine like winemakers and therefore, in judging their own blind, the results of the OCF were THE definitive evaluations of wine quality above and beyond all others. And – it was so perceived by many (most?).

As a winegrower I have, over the years, appreciated the results of the OCF greatly – maybe over most of the other judgings. I am probably professionally in-line with their sensibilities. I would, however, like to point out some drawbacks to that format of which we should be consciously aware.

Industrial winemakers tend to be very locked-in to their corporate identities and develop tastes exclusive to that to the exclusion of current experiences with other wines and art. In the business it is called "cellar-blind" and that means that they only find merit in their own product or style. Often they have little experience with other wines. A primary focus is that a wine is always the same regardless of vintage to appeal to the masses – no surprises. The small-winery artiste is often so deep in his/her personal expression of what is "possible" that they forget about "tasting good" or what the public likes. I'm sure we have all had at one time or another one of those table wines that is thick like syrup, butter in finish, tannic in the extreme, alcohol of 15 – 17% and tastes like an infusion of oak – liquid toothpicks, if you will, that put slivers in your tongue! And I have seen buttons almost popping off shirts with pride by the maker of such monstrosities! And by the purchaser of such a wine. It is beyond my comprehension but – different strokes for different folks! These types of winemakers are also judges at OCF.

Another aspect concerns "conditioning". Winemakers live with their wines – tasting them for their progress regularly. Hardly a day goes by in which a winemaker is not tasting *wines-in-process*. They become adjusted or conditioned to that situation – not the finished wines that, say, restaurateurs are exposed to. Not so very long ago, as white wines came to the fore in America, consumers would put a bottle to store in the refrigerator. Later, as they opened it, they would find what they called "ground glass" in the bottle and make irate calls to the winery and/or return it to the store. This didn't occur in a restaurant in the ice bucket – too short a time. This was not glass – it was potassium tartrates formed by the cold storage (simplified). The poetry of the old books called them "pearls of wine" and they are harmless. They were new to Americans. Wineries, to preclude this, began to "cold stabilize" white wines. The chilling of the wine in tank causes potassium to combine with tartaric acid and is easily filtered out prior to bottling. In fact, it is usually one of the very last steps. When the cold-stabilizing is done the reduction in tartaric acid results in a reduction of TA of about 0.1 or so (at least in our hands). Knowing that this is going to occur many winemakers carry wines-in-process about .1% *higher* in Total Acidity than what they desire the final marketable wine to have.

As a result most winemakers (I believe) become "conditioned" to higher acid wines than those that consumers are used to and like. This "conditioning" leads winemakers to appreciate wines higher in acid and thus to vote for those wines when judging. While I haven't paid attention in the last perhaps fifteen years or so at one time I did. The TA values of top whites at the OCF were higher than those of the LACF excluding common wines.

The wine competitions give us an independent "how goes it" relative to the changing public taste and knowledge. This has been of particular value to us at Ventana because so often we have been working on the future not the present. We have had to adjust to try to have the wines just on the outer edge of acceptance and not beyond. Bob Mondavi, in response to a question in a forum, answered that "Doug Meador is often ahead of his time and sometimes too far ahead". Cracked me up (I was there in Sarasota, Florida) because he was so serious!

Now that there is a plethora of judgings across America it has become popular among the pontificators to "put down" the judging results and medals. That is really a dis-service to the consumer. The phrase "Oh – everyone has a gold medal" just isn't true. In fact, judgings are far more rigorous today than in the past. The results are far more meaningful for the consumer particularly by region of judging. For example, the east coast has a far different palate then the west coast. Historically, the east has more of a European palate of long standing. They are closer to Europe in terms of transportation and have had those connections for more than 150 years. A judging wherein the judges are east-coasters will have results more in cultural tune with their eastern locale than would one from the west. This type of information can be a boon to a winery in guiding some of its marketing efforts. If you have a style more appreciated in the east – spend marketing efforts where it is appreciated. If California – then California, unless you are making "everyman's wine". In that case you probably don't have to worry about it anyway.

Another interesting thing about the judgings is that quality *does* show. In one aspect aberrations do occur because of the human element. If a wine receives a medal (gold even) at *one* judging and is never again heard from – it is probably an aberration unless it is from such a small lot that it doesn't make the quantity on-hand cut on other judgings. However, many wines repeatedly receive a medal at judging after judging across the country and across elapsed time. It doesn't matter if its gold, silver or bronze – those are human grouping things. But *any* medal judging after judging shows that that is a wine of merit and dependability for the consumer! I would personally have more confidence in a wine with 10 bronze medals than a wine with 1 gold then nothing.

One perennial problem with the judgings is the cultural prejudices of the judges. In spite of chief judges emphasizing over and over that wines are to be judged within their own varietal as to quality and merit *for that varietal* there are many judges that, for some reason or another, cannot get that point inside their heads! There are judges who believe that a white Zinfandel or Chenin Blanc or Green Hungarian should NEVER get a gold medal. Chief Judges have even resorted, at times, to shift the curve upward after all the scores are in just to compensate for that attitude.

One last commentary on judgings to aid the winery. Enter your samples as soon after the earliest entry date possible if shipping. Give them time to rest and recover from the vibrations of truck travel. Vibration is *very* damaging to wine. Years ago at the second Intervin International held in Toronto, Canada, Ventana had been awarded some ridiculous number of medals and they wished me to come to receive them at their awards banquet. I flew into Buffalo, met up with the chief judge who was also the chief judge of the California State Fair, and we drove together to Toronto. In the course of our conversations I alluded to this idea concerning vibration and how I would deliver wines personally in my Suburban carrying them *on* the back seat *upon* foam egg-carton type material. I would drive to Sacramento for his, to Pomona for the LACF and to Orange for the OCF. For Toronto I had shipped overnight delivery on the first day of entry. He thought that was crazy as hell! On the way *back* from Toronto he said he had thought about that craziness and decided that he would track delivery dates, methods (truck, etc.) and medals won at the upcoming California State Fair (by this time CSF had a commercial division). Not one medal was awarded, he later told me, to a wine that arrived by truck in the last ten days before entry cut-off! Amazing co-incidence if that is what it was.

Another aspect of this transportation is time of year and time of shipment. Always ship on Monday! If it is an out-of-state judging, ship on Monday next day delivery. Wines shipped late in the week may sit in sealed trucks over the weekend with temperatures climbing to 150 degrees plus if summertime. Wines have also frozen in winter! The protection of wines from climatic conditions *while* in *transit* is nil to below! Not so long ago they also weren't protected well once they had arrive at some judging's storage site. Those I delivered late but I myself delivered. Now I think most, if not all, are adequate.

At this writing The Ventana Vineyard is now *30 consecutive years* of gold medals awarded to wines from its Riesling grapes and *29 consecutive years* on its Chardonnay grapes from and including its first commercial crop. It has never missed. Wet year, dry year, hot year and cold – every single year! No single vineyard in the world has a string like that on even one variety much less two.

When I say "in the world" let me point out that there is a structural aspect to this. "Judging" is a relatively new phenomena – you can still see today European wines with a gold medal on the label from 1904 or such! Judgings in Europe were very rare and big-name labels don't participate – too much to lose. The statement "my wines aren't for judging.

They're for drinking" is still the common mantra. For an outstanding historical take on this read "The History of the 1855 Classifications" by Dewey Markham, Jr. – a tour de force work! Americans love a horserace! This regular judging is an American thing – probably original somewhat driven by our historical second place self-image and our striving to show we can be as good, even better, than our European mentors and standard setters. So – while "the world" has a nice ring to it, it really is within an American context. But – so what? It is our game so we get to make the rules. If you want to play enter your wines in our game! Simple.

As far as the red varieties are concerned it has taken a bit longer to be an overnight success – about 26 years to be exact. Remember, not so long ago the gurus were asserting that it was "impossible" to do reds in Monterey. One just has to be awed by the self-confidence (even when usually based upon incredible ignorance) of the pundits, scribes and critics that lead to such absolute commentary. Someone in Broadway once commented about critics that they reminded him of eunuchs in a harem – they knew what to do, they'd seen it done a thousand times – and they still can't do it themselves! Even here – the question isn't whether one can do it once but rather can the performance be repeated? Consistent high showing over time – year after year – is the hallmark of a great estate. A one-time thing may be in the "even a blind pig can find a truffle once in awhile" type thing. The Syrah golds started flowing in the 90's including a gold and silvers in Paris. Pinot Noir had been generating medals with sporadic golds since 1979. Now there is quite the rave about the Santa Lucia Highlands – rightfully so – but Ventana of the Arroyo Seco district (located just below the Santa Lucia Highlands) has been there long before – and continues to perform to this day. Grenache Noir is now coming into its recognition garnering golds in recent judgings. The grape family most derided from Monterey is Cabernet. These grapes, after sporadic fits of golds and silvers depending upon the year, are now gathering golds, silvers and bronzes more steadily. The recent Chronicle judging in San Fran gave gold to both the Ventana Cabernet and the Meador Estate Magnus – a Bordeaux blend. The years of work on Tempranillo are coming to fruition with that red varietal garnering medals. Merlot has been the recipient of many medals however now the small amount grown is used in blending. Specific clonal work on the Merlot has yet to come to fruition in decent quantities. That is ahead. Sangiovese in conjunction with Cabernet Sauvignon was the number one blended red-wine for America a few years back and continues to garner multiple awards (VV – "Due Amici").

All in all, it is my opinion that the wine competitions have done tremendous good. Sometimes the results have been flaky, sometimes bizarre, sometimes this or that. But those results have never been a function of how much advertising has been placed or of a paid trip to Europe or Australia or a fancy dinner and show tickets in Manhattan. Never. There are too many "judges" to buy. Within the industry we should all encourage the multiple judgings – so much comes from them that is socially desirable. They are the vehicles that some young artist, say in Kansas or wherever, will come to our attention. Some new technique will be brought forth. Some new dazzling region will emerge. Will the vested interests like that? No. In fact, they will try to tell you that the medals are meaningless. But our society will damn-well benefit. And so will the consumer's table!

With all their flaws and differences, these judgings are a lot like Democracy – a lousy system but what, exactly, is the alternative? Tyranny? Well, we have a bit of that bitter pill in our system also.

Tyranny is the exercise of Power beyond Right.
—John Locke

It is not the critic who counts; not the man who points out how the strong man stumbles, or where the doer of deeds could have done better. The credit belongs to the man who is actually in the arena.

—Theodore Roosevelt
Speech at the Sorbonne, Paris

Section 5
GURUS, SCRIBES & PUNDITS

Guru (1): A spiritual teacher, esp. one who imparts initiation: elder, teacher

Scribe (1): informal often humorous; a writer, esp. a journalist

Pundit (1): an expert in a particular subject or field who is frequently called on to give opinions about it to the public. Origin; Sanskrit "pandita" learned

—*The New Oxford American Dictionary*

This is a touchy subject but one that I feel must be addressed nonetheless. Over the years my relationship with the media has been stormy in the best of times and non-existent, essentially, in the rest. My naiveté was a hindrance initially. As the change in American culture was occurring and we were learning so much so fast about food and wines I thought part of my job was to teach others these newfound revelations. I thought the scribes' job was to carry the messages to the consumer – bringing truth and debunking myth. That service involved study and learning and diligence. Alas, in most (not all) cases it was not to be. It was far easier to fill one's rice bowl and more rewarding to periodically re-gurgitate old myths and legends as truths. The myths and legends were "comfort food" and sparked no real controversy.

I title this section as I have because it appears to me that over the last thirty-five years of my experience various writers fit generally into one of these categories more than the others – at least as to how they seem to fancy themselves. I say "fancy" themselves because the wine "reporters" to the public are one and all self-appointed. There is no school for wine reporters. There is no degree in the subject nor is there any certification exam or requirement to demonstrate any level of competency in wine or viticulture to anyone whatsoever! It is a wide open laissez-faire endeavor which carries with it lots of free wine (read: samples for "evaluation"), free dinners (read free dinners in expensive formats paid for by wineries), free junkets to far away places (read research and familiarizations paid for by wineries), free tickets to events, and often paid participation as a judge (read more free wine – after hours. The wine of the judging is a burden not a perk). The power of the pen and access to the media format as a gatekeeper allows the assumption of authority far beyond that which the merits would warrant.

There are among them the simple workmen – the scribes. They are simply doing a job of filling a certain number of inches on a timely basis meeting their deadlines steadily and making no waves. They take their paycheck at the end of the day and make no particular claims of superior knowledge or connection with the gods of wine concerning whatever. Among these people we see no mentionable pomposity, arrogance, snobbism, holier-than-thou-ness, or other really unpleasant traits. Okay – maybe smidgens, but not much. They are reporting on what they have tasted in that time frame and generally stay, rightfully, in the "I like it, I don't like it" zone.

One of such that comes to mind, and for whom I had great respect not just because of the above, was Jerry Mead, now deceased. Jerry just had a love of wine and he really liked people who like wine. He formed the quite successful at the time WINO clubs and he headed and helped form the Orange County Fair, Orange County Wine Society and New World International. With Jerry – over dinners or at events – one could have great discussions or even arguments, explore new observations in the vineyards or in winemaking and just generally get into contentious areas but when it came time for writing his syndicated newspaper wine column he usually confined himself to a little discussion of the winery then a very personal evaluation of the wines based upon his hedonistic response to the wine. At other times he was reporting on the results of his competitions or WINO Club evaluations. If you were a person who had found compatibility with his palate and opinion then he was a very consistent pen to follow. He and his type did a good service to the public.

Another of this ilk, perhaps with a little more showmanship and a shade more difficult at times was Robert Lawrence Balzar of Los Angeles. I think Robert was an older man at the recording of the last ice age – he may even have had a small dinosaur as a childhood pet! I think the world of Robert. He was promoting and writing about California wines long before it became fashionable to do so. He at one time had a wine shop called Balzar's in Hollywood – long before I was here. I knew him as a writer for the Los Angeles Times, newsletter writer and restaurant reviewer having first met him in 1979 in Santa Monica. Robert literally lived for promoting and teaching about California wines. Robert really built the foundation for the wine columnists that came later. Robert's LA Times weekend magazine sketches of wineries, their people, the vineyards associated therewith really was the seminal format utilized in various forms by subsequent practitioners. In the later years when his advancing age led to a strong decline in his media influence and then a cessation the respect and thankfulness for his earlier work would be displayed by many winemakers. Robert conducted wine classes in Long Beach and Los Angeles –

The New Viticulture

usually Monday in Long Beach and Tuesday night in L.A. or Lawry's in Glendale. At each of those nights, there would be owners and winemakers in attendance showing the wines of their region. The last fifteen years or more he had no pen so there were no marketing reasons to be there just respect. The winemakers and owners were a who's who of the industry. I haven't been in the last five years or so but Reggie Hammond has been representing me. I've made sure he knows why and he does. He says its actually fun to watch a late-nineties guy so excited about wine you'd think he'd just discovered it!

These types of wine scribes were/are very beneficial for the wine industry helping the consumer navigate the ever-increasing confusion of labels, much of which is just "plonk" – hookers dressed up in marketing clothes for the unknowing.

The "Gurus' are a whole different situation and really function more as a cult. Their palates have become divine to their followers, every utterance to be worshipped and woe behold he who fails to kneel at the altar. These are one-man shows and the egos that develop are incredible. Sadly, that power can have, and has had, significant downsides to it. I have no argument with a palate having faithful followers but when it is one person's palate only and that control becomes too large disaster follows, particularly if there are any criteria other than the taste of the wine involved.

As I have commented, the wine scene has been undergoing incredible changes world-wide over the last 25 years or so – more changes than have occurred in the last thousand years probably. Yet, often the palate of a single person can remain fairly constant particularly as the person becomes older. We older people tend to get locked in to our likes/dislikes and prejudices. If the media vehicle has an aged Guru of strong self-belief it is entirely possible (probable) that many new innovations are not only overlooked but that winemakers are overly focused upon a benign smile from the wise one that they are not even innovating or evolving as they should. Times change. Populaces become more knowledgeable. Gurus are locked in the past. If they've said something or printed something – like academia – they find it ego-wrenching to change or admit error. It is not possible for a Guru to be in error and when it appears that they were on some facet it damn-well was not their fault!

Let me give you an example of what I speak. You, by now, are aware that I think high alcohol and heavy oak are a faddish gross misdirection that our industry took for many years. One Guru publicly, and often over the years, stated that he liked high alcohol and high oak. Now – there is nothing wrong about that – I think the 1st Amendment says something about that. But when one is in a position of affecting the social *geist* then one should have some sense of responsibility. I am of the opinion that that Guru strongly influenced that misdirection and impedes its demise today!

When a new area or region comes into being there are struggles in its birth. If such a region acquires the disdain of a Guru then the memory of the elephant takes over and infallibility raises its ugly head. Never will the Guru change course even if truly knowledgeable folks come to some opposite consensus. To quote "It is impossible for Monterey County to make adequate red wines" end quote. Amazing! Even the word "impossible" implies some sort of superior divine knowledge. Here Monterey is some twenty years later and a mountain of gold medals (even from Paris!) on red (yes – believe it or not) wine and the attitude is well, those are just those silly wine competitions! This scenario is not just Monterey but the experience of other emerging areas as well. I'm just more cognizant of Monterey's experiences.

Another Guru has encouraged the making of over-the-top over-extracted unbalanced high alcohol wines by bombastic use of superlative words all geared to bigger is better couched in almost legalistic phraseology in its conciseness and crispness. In pandering for even the slightest recognition of existence from the god who walks on earth, winemakers and owners push hard for a style of wine to please the deity and ignore that which pleases most other people. The really funny thing, to me, is that I know this Guru's palate and he does not really find favor in these sorts of efforts. He does have limits within which quality and merit lies but his language leads those who do not possess his internal artistic limits and comprehensions to continue to go beyond the pale. Sad. On the up side I do see more and more young winegrowers saying "To hell with that stuff. I'm going to do my own thing". To this I say – "Fantastic, youngster – show us what drum you hear and dazzle us. We're waiting with anticipation. Take us into the next level!" I've had my turn in the arena. Now I want to cheer for you!

The third of my categories are the "Pundits". These are the "learned" ones, the ones who have devoted the time and energy to acquire as much recorded knowledge as they can on the/all subjects. Often these strike me as more involved with one-upmanship of "knowing" than of true love of wine. The writings of these sorts are not really the type to stir the passions of the reader for wine but rather to dazzle themselves with a display of their erudition. Often the pundit's writing is such a convoluted presentation that the layman's or consumer's attention is soon lost – if they give a damn in the first place. Do we really care that a wine reminded one of the failure in Hanover and Leipzig of Brahms's Piano Concerto No. 1 in D minor? Even if it is true?

I like to read these folks. Sometimes they will say something of interest but usually I am just finding humor in the incredible seriousness of it all. It is interesting to point out errors in a technical treatise to them and observe how it is taken. A true seeker of knowledge appreciates accurate correction. A poseur on an ego trip doesn't!

As you can no doubt tell I am not a big fan of the writers and critics. But then, show me anyone who actually goes in "harm's way", goes into the arena and faces the tiger, puts their work and their ego up to the public that actually likes having noncombatants give their opinion – be it a stage play, a wine, a painting or a sculpture. When you are at the point of the spear all alone life is so great – and the critics, sadly, will never experience that.

You young winegrowers – prepare yourselves. Learn and practice under a mentor if you can (it's faster) or on your own if you must. Then go into that arena, face that tiger, take your wounds, heal and rest then rise and fight again. Go back into the arena and ignore those sitting on their butts safe in the stands. You *are* the point of the spear! To the best of my knowledge not one of the critics *has ever gone* into the commercial arena!

I do encourage one to read the critics just for another "feel" but don't let the downers depress one and, for God's sake, don't believe the upside, the high praise. That's for what you did yesterday. You should be looking at tomorrow.

I recommend that a winegrower spend as much time as he/she has reasonably available in the tasting room *serving* and *interacting* with the public. Don't get too caught up in their comments about the wines as they will be mostly plauditory – after all, they are in your house *and* are being blessed by the presence of the great one. Do pay some attention to any negative comments. However, those and marketing are not the reason you should be there. The reason is *body language* for it is there truth lies. Spoken language is filtered. Body language is truthful to a much greater degree. There is where you will find what you need to know. Finding out the "why" takes a little more work.

Our mission as winemakers is to provide pleasure. Period. Looking at people will tell one where one has to go.

Beware of taking any one thing out of its
connections for that way folly lies.
—Ralph Waldo Emerson

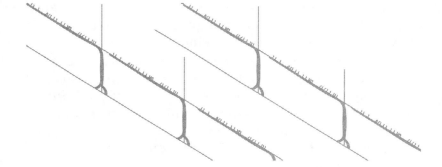

Academism results when the reasons for the rule change, but not the rule.
— Igor Stravinsky

Section 6
UNIVERSITIES

As a winegrower, researcher and employer I have some very strong objections concerning the quality and comprehension of the products the schools are tendering to the industry. The students themselves are not at fault – they are, in general, sharp young people who have simply done what they were instructed to do along the way picking up entrenched biases to their detriment. They come to us in the industry possessing their new bachelors degree and are sure they are up to date on the rudimentary knowledge of their chosen profession because the schools have told them so. They are sadly mistaken.

First and foremost in the damage hall of shame has been our allowing – historically – academic elitism to creep into and distort our basically agricultural discipline at the undergraduate level. It has been the habit of western culture in general and America in particular to look at tillers of the earth with disdain. This historical view of peasants and serfs (there is a distinction believe it or not) as brutish workers in the fields meant by God to be ruled by the nobility and the educated has permeated to the present time even though conditions changed long ago. The noble class has gotten the message (at times forcefully) that they are an irrelevancy in modern times. The education elite have sprung forward with great vigor to fill that ruling class void – as they say, Mother Nature hates a vacuum! With the demise of the secular power of the Catholic Church the education community has again sprung forth to become the new priests guarding accumulated truths that they venerate as strongly as the old ones did theirs.

As time has passed we have seen gradations develop within the education elite – a pecking order if you will – often based not upon merit or performance but upon pedigree. How often have we seen the smug, self-satisfied announcement of Harvard or Princeton pedigree in the company of Berkeley or USC people? The gradations go even further. Ever watch a physicist or mathematician when someone announces their sociology education? These stratifications permeate the class. At the bottom of the totem pole lie the hands-on disciplines – primarily agriculture related. The curious aspect of this is that the very existence of the wealth that allows the education elite their sedentary and contemplative life of ease flows *from* the bottom of that prestige totem pole.

Throughout history the existence and size of the leisure class has been a function of food production surplus to the society's needs. To the extent that food production sags below those needs the privileged class disappears only to rise again in times of surplus. Look carefully at the time frame from 1348 onward. That year the bubonic plague hit Western Europe resulting in death of around half the population. There was an extreme shortage of workers for the fields. Many of the rules of centuries standing fell by the wayside. Serfs abandoned their Lords and ran off. Peasants and serfs alike demanded new rights and for higher cash wages for their labors. And they got them. Many of those new gained rights in England are still fundamentally codified in our laws today. It was long ago but we still are influenced in our daily lives by those events – not the least of which is the freedom to move about and seek wages from the highest offer. Another right gained was the right of peasant farmers to plant crops as they saw fit in response to a market or needs of mother nature (replenishment of the soil, etc, which they *did* know of) rather than repeatedly plant the same crop under orders of their noble. The soil will tell the farmer the crop and the best way to do it.

Today, this "dumb farmer" has just about every aspect of modern life focused on him or her. Consider the following subjects with which an *individual* farmer has to be very knowledgeable in order to survive within the People's Republic of California: Immigration Law; Labor Law; OSHA; Finance; Marketing; Accounting; Environment Law; Politics; Agricultural Chemicals; Fuel Laws; Equipment specifications and applications; Insurance; real estate and the list goes on and on until we *finally* allow him or her some time to FARM a crop! The "dumb farmer" routine is ancient history. Most of our modern farmers are college graduates of one sort or another.

For a very long time (perhaps from the beginning until the present) our discipline has been best characterized by the concept of "Winegrower". That is, a man of the soil grew the grapes each year, harvested them and turned them into wine usually all on the very same property. Subsequently any that was excess to family needs was sold or traded to merchants who specialized in transport and marketing. The merchants were not winemakers then nor are they today. The winegrowers, as peasants or serfs, were people of the soil seldom traveling more than 3 kilometers from their spot of birth. They received no formal academic education historically. That is not the same thing as saying they were not "educated" nor is it saying that they were stupid. Far from it. Try being stupid and surviving on what you grow. If you failed there was no unemployment pay to bail you out. Starvation was the penalty for "stupid". How many of you today can start a fire with wet wood at 30 degrees below zero with no matches and with just a leather thong? How many can predict the coming weather by "feeling" the air? Can you weave a blanket or clothes to keep you warm? Or even make thread to weave? Do any of us know which wild plants are safe to eat or which to use to cure sickness? This *closeness* to the soil led to dictums based upon close observations of natural phenomena. That is the inductive method – observe then generalize. Some argue that the deductive is the better tool (as did Einstein) but I ask where did the accepted eternal principles come from if not induction? Well – that's an unimportant discussion here.

Just as various fruits and vegetables have been hybridized for *travel* and *shelf-life* without regard for *taste* so viticulture procedures were recommended by academia without regard for the glass of wine. These people were not malicious. I think, rather, that they were poorly prepared by the academic system within which they had been molded.

The point of all this is to assert that the growing of grapes and the making of wine are merely stages in one *continuous* process with no one part of more or less importance than another. The glass of wine is a summation of each and every micro-step along the way! One separates these at great peril! Our ancestors (and many still today) had a built-in loop feedback system wherein errors could be readily identified. Faults in a grape were very apparent in the wine. One may argue that how could one tell because the wines were, in general, so bad to begin with by our modern standards. I say that if "bad" was the standard then there were still many different levels and types of "bad".

In support of my assertions I do point out the organization of most famous-for-quality operations. The great chateaux and estates throughout the world are integrated. There is no cleavage between growing and winemaking and the results speak for themselves – they are "famous" to some degree or another. The loop feed-back is in place and operating.

The California schools of today either do not understand that winegrowing is a continuum *or* have deliberately separated growing and winemaking at the undergraduate level from academic totem-pole hubris to the great dis-service of the students and society! This elevation of the "scientific" number-crunchers over and above (and thus separate from) the practitioners of growing the grape has resulted in a cleavage between the two parties such that they don't speak a common language. Absent a common language and cultural foundation how can they possibly communicate?

When contact does occur it is in the form of a Lord dictating to the serf. Usually the Lord has no idea of which he speaks. Some of the most ignorant folks I have run up against have been winery field reps. They, in general, are totally without a clue about what goes on in a vineyard. Oh – they pose nicely enough and I'm sure they impress their corporate bosses but not the growers with whom they interact. Don't take my word for this – ask around for yourselves.

> *What is any established institution but a society*
> *for the prevention of change?*
> —Lewis Mumford

The scientist winemaker speaks a chemical language and his mind works in that fashion – solving situations with chemicals. The grower speaks a botany language, a soil language and a weather language. A grower looks to solve a problem in those terms – not chemistry.

With the structural inability to communicate the connection of it all is lost. At the industrial wine level it has been lost for a very long time now. In response to this loss we have seen the advent of very large "farm management companies" who *do* speak the *corporate* language well with the industrial producers, who make it easy for one-stop shopping, who develop personal "rapport" with their corporate counterparts and have no interest *what-so-ever* in the outcome other than did the grapes meet contract specifications. We now abound in these operations and their acreage management portfolios are huge.

When these grapes in their train-long line of trucks arrive at the winery they are assembled into monstrous tanks – some a large as 600,000 gallons in size. Where, exactly, is the loop-feedback? How would a grower know if there was a solvable problem in his fruit or if his was superior and covering up for weaknesses in other grapes? No feed back exists. No road signs for improvements – just blind acceptance of the status quo is the result of this system. The results are in the hands of modern chemistry and the practitioners of that arcane art. How can one make progress in improving America's wines under this scenario? This system penalizes the superior grape and rewards the inferior – one size fits all.

Not too many years ago the executive winemaker of a very large industrial winery announced that they were embarking upon a program to evaluate the wines from each grower. The good ones might be paid more and the lesser ones less or be dropped as suppliers (notice "suppliers" language not "growers"). There were perhaps a hundred growers at this gathering. I thought at first that this announcement surely was just some B.S. for the marketing folks. But it wasn't. They actually believed that (a) they could do such a thing and (b) that their self generated information was relevant. This was also at a time when they had acquired ownership of a huge amount of vineyard acreage!

Keep in mind that said winery didn't have the facilities to crush separately and ferment each grower's grapes. Secondly the presses they used were not used by any fine wine producer in the world of whom I was aware. Thirdly, which yeasts were they going to use and under which conditions would they ferment – slow or fast? Fourthly, fifthly, etc., - list goes on.

These industrial winemakers were, apparently, going to try to follow small lot fine winemaking procedures with large lot knowledge and equipment. Boy – that is really a full plate. Just the recording and reporting would be something not to mention harvest scheduling. I was awed by the magnitude of their undertaking and the hubris associated therewith. Noting their newly acquired vineyards I also contracted my one lot of those grapes contracted to them to another party effective when the existing contract expired in one year. There was no option clause to worry about.

One very good example of the effects of this cleavage comes to mind and in which I was a participant. Once upon a time there was a grape fairly widely grown that was called the "Napa Gamay". U.C. Davis and the FPMS declared, repeatedly and continuously, that this grape was the Gamay of Beaujolais. The wine was so labeled as such – Beaujolais or Gamay Beaujolais. The fact that the wine therefrom tasted nothing like the Beaujolais of France seemed to bother no one (until me) and Frenchmen who tasted the wine were reported as saying "Of course it doesn't – you do not have the soil" and we Americans acquiesced in that then current dogma. I was of the opinion then (and still am) that soil can/does have influence in nuances but that winegrowing procedures cause far, far greater effects such that soil nuances will only show with time and within a relatively fixed farming regime. Bill Jekel and Bob Mondavi had some widely reported discussions on this on-going opinion-fest during that very time – a fest into which I was drug from time to time by each side even though I tried very hard to avoid the hassle. My plate was full and at that time I felt we didn't have enough dependable information.

I had some Napa Gamay on my vineyard and I was very confused about this thing. I knew what the "experts" were telling me but it made no sense whatsoever. Now, in Beaujolais, the Gamay grape (properly known as the Gamay au Jus Blanc to distinguish it from the other members of the Gamay family which are all tinturiers – colored juice – to some degree or another) is an early ripener. In fact, for a hundred years or so, in Paris there is a Beaujolais celebration of a successful harvest held historically on November 21st of the harvest year. Wine in barrels was delivered to Paris for the party. Today it is in bottles and makes its way also to New York City and elsewhere by air.

Given those background essentials I looked at THE WHOLE PROCESS! The Napa Gamay grape grown in far warmer areas than the Beaujolois region of France rise to equivalent sugars much *later* than they do in Beaujolais and acquire darkened seeds much later! Further, following the Beaujolais procedures of maceration carbonique (a slow fermentation technique compared to massive industrial starter approaches) it WAS NOT POSSIBLE to ripen, harvest, ferment, M-L, settle, clarify (no centrifuges then) stabilize, bottle and ship the Napa Gamay grape by November! Conclusion – it was not the same grape! It might also explain the fact that it didn't taste like Beaujolais. Rudy Neja, then the viticulture extension agent for Monterey, upon my discussion with him sent a letter to a French vine identification team then working in Australia. Australia had all our FPMS materials. Rudy asked the French team to identify that grape for me. They did. It is the Valdiguie.

Jerry Mead took it public in one of his columns. The Vit Extension guy in Napa at first said "yah – we knew that" but subsequently was "no comment" on it. You see – if they *did* know and continued *selling* plant material falsely – that's a legal problem. I think they did not know and were just trying to cover an embarrassment of incompetence. You see, here all that was required was a knowledge of BOTH viticulture AND winemaking combined with a skepticism of authority when observations or facts indicated otherwise.

In my opinion there should be *one* undergraduate degree for our industry – Bachelor of Viniculture – not viticulture or enology or fermentation science. Students should be indoctrinated thoroughly in the "continuum" concept of their profession. At the Bachelor's level they are being taught "tradecraft" and simply that! Specialization should be at the Masters or PhD level if students' interests should take him/her in that direction. The course of study should include *at least* one full intense year of general plant physiology. More would be better. Give me a kid who understands that and I'll teach him about a grapevine in – oh, say about a week!

Fermentation fundamentals at the practical level should be taught to all including all used product tests and the equipment used therefore.

Soil science of one year should be taught as a minimum and the use of additives normal to the industry thoroughly explored – both pluses and minuses.

Geography of the wine world and the grape varieties associated with regions and soils should be taught. I can't believe the number of young *viticulture graduates* who haven't a clue about grape regions of the world or varieties grown in different areas.

Agriculture business management should get two years attention and include accounting, finance and marketing courses.

Practical Viticulture science should include planting, light, pruning, training, harvesting, etc. Particular focus should be placed on the relationship between a viticulture action and its purpose in the end *wine*. In the fourth year, once the foundation is built, there should be a focus on solving wine problems with the vineyard practices.

There should be some exposure to art and music in an "appreciation" form. After all, art is a component – an important one – in their profession. Art and music have a funny way of rubbing off the rough edges of the barbarian and the young – same thing, I suppose.

English *and* Spanish should be a requirement – at least at some fundamental level. Speaking Spanish is simply a "must" in California today – particularly in our world of wine. It would be nice if the students could read and write adequately in English. I really like to see sentences in applications with verbs and nouns spelled and organized such that I can tell which is which.

There should be a class on interviewing and application presentation. Many years ago I interviewed three recent graduates of U.C. Davis for an opening for "winemaker" – a term we don't use at Ventana. Each one of these incredibly arrogant snots proceeded to tell this grower how he was going to "make" my wine and how he would "improve" them. Not one asked what I had in mind. Not one had even tasted one of my wines! Needless to say I hired not one of them. Both subsequent people in that role have trained for several years *first* in the *vineyard*, then the winery and then supervised both. Both speak Spanish and one also German plus a smattering of French. Both are able to lecture at length on wine regions of the world and the grapes grown therein as well as the prevalent winemaking and growing techniques associated therewith. In other words they are "professionals" in this worldly discipline.

The observations of all the "specialists" working in their respective little worlds raises the question "Where the hell are the Generalists"? There appears to be no one (not one person!) in *all* of *academia* whose assigned mission is to put all these isolated little revelations into a system, conduct a "systems analysis" (the application of which I brought to their discipline from aerospace where it is the norm), evaluate relevance and meaning and tender it to the practitioners in the field. Without an integrated format or system the specialists don't truly have any idea of the importance of the pieces upon which they are working. Usually the motivation is just something that has caught their interest and /or that they can sell to a funding committee.

A fellow by the name of Erwin Chargoff once asserted that science is wonderfully equipped to answer the question "How?" but it gets terribly confused when you ask the question "why?" The chemists of wine have done a remarkably good job of identifying long lists of molecules and compounds of various types and sizes and done well on how the chemical processes sequence. As chemists they focus and rely upon chemical solutions to problems or questions that may arise. From the short-run industrial process point of view they are probably correct in doing so. After all, industrial level conditions are such that processors must deal with the mundane responses to base resources as they present themselves at the moment. In an industry such as ours the basic raw materials are generated once a year but the supply of inputs (bulk wine) may appear at the factory at various times from a wide array of sources – both domestic and foreign. The wine chemist's role is to determine "what to do" to process the material not to necessarily dwell upon "why" it is in the condition it is. Turn-around time is of the essence. The industrial focus is that of the shortest time necessary to achieve the widest possible acceptance at a given, profitable price point.

I really do question, though, the contributions of the academic chemists even to modern industrial wines. In a small winery such as Ventana the winemaking practices haven't changed appreciably in twenty-five years. There is some minor "tweaking" in yeast choice but that's about it. Yet – the medals at the judgings keep rolling in. I know we don't pay any attention to Davis' work. As I wander other people's wineries I don't see any expression of technique acquired from there. Yes – I see tanks being filled with lumber to simulate oak but I doubt that came from the schools. The spinning cone didn't come from there. The centrifuge didn't come from there. Etc., etc.,. Does anyone know just what they do and to what purpose? How and on what are they spending our money? Besides teaching? There is a difference between "putzing" and working towards a purpose of value. Without a direction, a system as guide, a goal how does one know if one is putzing or not? Further, how does one even know if there is a problem and the *need* to solve it? There is a finite limit to the assets we as an industry and a society have to devote to research and these orders of priority. Do we wish to fund "putzers" or should we demand responsiveness to our pressing questions? Years ago, I sat on the Wine Institute's Technical committee. In this case the Institute was funding research projects. Annually, the academics, if they wished monies, came to the committee, presented the results of their previously funded projects, presented their proposed future projects and the purpose therefore and requested funding.

The answer was yes or no. Some of the work was good, some not. The message was fairly clear – give us what we need. When the funding flows from amorphous habitual structural source the accountability aspect is lost. We, as a society, are simply funding a gentle life-style not demanding production of timely and useful knowledge in a structured format.

"Putzing" itself should not be denigrated. Putzing has yielded many fine advances and revelations. But – putzing is something to be done on one's own time or after the main work is accomplished – not in lieu of the main work.

There are a lot of industrious people in the academic enterprise that develop all sorts of interesting little "facts", bits and pieces scattered all over the place and there they be! There was a time when U.C. Davis didn't even look at Fresno State's work and vice versa! For example, at the same time that Professor James Cook, et al of U.C. Davis was discarding foliage feeding, the effects of which were too minimal to continue in their study, Professor Vince Petrucci of Fresno State was reporting that yields were increased with foliage feeding but maturity levels were not affected therewith (Petrucci & Thorsen: V-102-80)!

Can anyone imagine building a skyscraper with no architect – just concrete engineers who do not speak to the steel engineers who do not speak to the electrical engineers, etc. – all going to work and doing whatever strikes them that day? Well – that is exactly what has been going on today in our industry! The very people who are tasked with training our young practitioner do NOT train them in an integrated system approach and then somehow expect these working folks in the field to later sift through the gems of knowledge and assemble a working whole.

I also point out that these little gems are developed at the advanced degree levels and are presented in language and nomenclature that is somewhat foreign to the every day person such that even if the "gem" were germane the obviousness of its importance is easily missed.

I make the point that "science" is *organized* knowledge. "Science is built up of facts, as a house is built of stones; but an accumulation of facts is no more a science than a heap of stones is a house." (Henri Poincare).

Besides teaching, the primary functions of universities are to discover "facts" AND the organization of those facts into meaningful structures. When these "meaningful structures" will not contain observable "facts" then either the structure is faulty or the "facts" are. These aberrants will automatically guide the explorer into properly questioning the fact or the structure.

The production of "facts" for the sake of accumulating "facts" is a penchant of the chemistry side of our wine academia. The rewards associated with the generation and publication of unstructured "facts" are those related only to the *academia profession* and are not of any general use to the wine profession! *Action* is the proper fruit of knowledge. That "action" by academics is the application of generated knowledge to a structure and the resultant continuous review and evolution of the subject structure. Academia mustn't drown us in details – it must look at the whole. We pay them to think about things of value to *our* profession – not to pad their curricular vitae. Academia has expanded our knowledge tremendously of minutia and detail but not our capacity to use them with wisdom.

"The test of real and vigorous thinking which ascertains truths instead of dreaming dreams is the successful application to practice". (John Stuart Mill). That sentiment is every bit as profound today as it was of yore. It probably is one of those "eternals". "The actions of men are the best interpretations of their thoughts". (John Locke) is another.

For perhaps five hundred years gunpowder was merely a Chinese curiosity used sparingly for entertainment because, until western minds came upon it, it was not put into a structure of thought. There are, similarly, literally volumes of "facts" about which nobody knows or cares because they apparently have no relevance to what we do. How, exactly, do they fit into our required operational structure? Do they support or deny our methodology? If deny, is it the methodology? Or are they simply irrelevant to our present mental capabilities and to be set aside until such time as their existence becomes relevant?

On the viticulture side of our profession we have been somewhat derelict as scientists on the chemistry of the vine because of our failure to develop a "structure" that would demand it. If we had a structure or system of viticulture to which all questions or answers must be directed and compared then the need for this work would have become apparent long ago. If we as a species can identify, name and list every arcane molecule in wine then we should certainly be able to identify every molecule or compound in a vine's sap at the various times of the year or cycle. We should know every hormone in a vine and exactly what its function is. Why? Because those hormones *control* what goes on inside the vine. Their inter-relationships with nutrition and temperature should be mapped. This is, and has been for a long time, within our capabilities! Yet – today we do not know these things! Our researchers are chasing piss-ants while its raining elephants. In order to construct the very best systems in given areas we must know "how" the vine works and "why" it does what it does. Only then can we – with confidence – intervene and exert control to our ends.

Thus, I suggest that there be, in this new recommended format, a Professor of Viniculture at the graduate level only whose function is the assemblage of information into a viniculture system AND the continual exploration of potential new systems in light of newly developed knowledge. This "chair" should be the superior seat in the hierarchy – it is the ultimate purpose of all the other more minor activities. This "chair" should be plural in the sense that no one person should rule and dictate – which is what we have seen in certain areas in the past. Members of the panel composition of this "chair" should be of proven competence, tenured but time limited, and each allowed to be "rainmakers". Biannual performance, evaluation and review analysis of existing systems and any proposed new ones should be disseminated to the industry and the various trade organizations. These reports should be signed.

These are the things we employers need to have – all the professional trappings without that damn cleavage that has hurt our industry for so long.

I implore the universities to address this issue. I'm sure that turf wars would ensue in pursuit of this course – the vested interests are strong. But if any sense of social responsibility exists within academia it should be done – by dictate if not by consensus.

Or, perhaps, a school that wishes to make a mark for itself that does not have the entire deadwood encumbering itself would follow the format. It would quickly become the "darling" and its graduates would see open arms. Cal-Poly, are you listening? Oregon and Washington State – what about you?

To truly understand the methods and philosophy of folks don't listen to their words, fix your attention on their deeds.

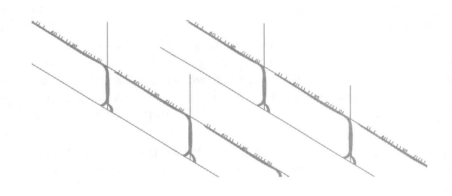

*Ideas or Hypothesis are tested by the consequences
which they produce when they are acted upon.*
—John Dewey

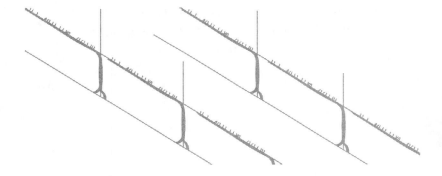

Discovery begins in Wonder.
—JDM

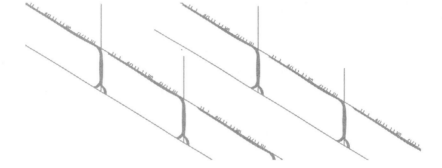

The smallest details of this world derive infinite significance from their relation to an unseen divine order.
—William James

Section 7
PICTURES

THEN

1972

NOW

PLATE 4: 2/08

The old standard UC Davis system of 12 feet by 7 feet still exists in many vineyards today. The conversion to the new viticulture is not yet complete. This picture was taken in 2008 in windy Monterey. Notice the length of the Chardonnay canes. There are few over 24 inches with many much shorter. Also notice the extremely short internodes demonstrating the struggles for growth the preceding summer. You might also notice, looking down the row, the preponderance of growth material of the left hand row compared to the right hand row. That left hand visual is the down wind side. During the growing season the wind piles up the foliage on the south side. Look at the unused land space down the center of the view. One can see that no "solar radiation receptors" (leaves) occupied the area.

PLATE 5

Here we see the vine system on the Ventana many years ago. There are two lines of vines where before there was one. The rows are 6 feet apart. The cordon wire is much, much lower. The vines are 3.5 feet apart within the row. The previous picture indicates that within the farming regime and climate they cannot generate enough vigor at seven foot spacing. The closer spacing allows the vine to generate sufficient energy. Both pictured vineyards are on their own roots.

Simple arithmetic tells us that two rows where one was before should result in a doubling of yield per unit area. The form is unilateral, unidirectional. This was early in its development and it had not yet graduated to heavy use of the second wire which would further increase yield. The design side view can be seen in the April, 1986 article in the Wine Spectator.

PLATE 6

This was taken sometime in the mid-eighties when the conversion to the "third" row – as described in the text – and the interplanting of the original rows one and two at the 3.5 foot position. This was very early on as one can see the use of wooden Keruing 2 x 2 end stakes. We still had not solved the most desirable end post format.

PLATE 7

Here we see the insertion of the 4th row and the use of the new end post design. This was in 1990 when I had all the equipment design problems solved and we moved ahead with the 4th row installation. Terrell West designed the post with Quiedan's assistance and involvement. Dan McNamara considerably improved the concept. The Ventana was the first place they were used. At the time they were incredibly inexpensive as they utilized used oil field "stinger rod" used in drilling. To the oil business they were scrap that just accumulated. We could get them for scrap prices. Dan cut them and welded on the resistance plate. Stinger rod is very thick-walled pipe thus strong. Later on, the popularity of the post eventually eliminated the piles of scrap rod. Manufacturing the posts from new material substantially increased the costs. The posts were simply driven into the ground to proper depth by a hydraulic tractor – mounted post driver. Quick, inexpensive, long-lasting and strong!

PLATE 8

The Ventana some years back with this Chardonnay fully developed in our new fashion of 3.5 by 6. By looking carefully one can see the younger vines in between the older original vines. The movable wires are still up from the previous year's growing season. Soon they would be taken down and tied at the base of the vine for pruning not to be raised until the needed time the following season.

Compare, in your mind, the radical increase in exposed leaf surface here versus that of picture four – the old system.

PLATE 9

I include these pictures to display my comments in the text concerning trellis. These were taken in 2008 in a coastal vineyard quite famous for the quality of the wines made from these vines. Vineyards such as this are becoming more rare with the passage of time. You can see in the background the wide spacing given to these "free standing" unirrigated vines. Vines grown in this fashion are still common in the plains of La Mancha (in Spain – Valdepenas) and in the south of France. You can also see them in the Lodi area of California and a few scattered in other areas of California.

The important aspect I wish to show here is the penetration of light deep into the vine. The arrow shows the direction of light penetration and the canes from the previous season bend out and away enabling that penetration. The fruit receives diffused light but the leaf distribution protects from sunburn. This system is not amenable to machine harvesting – it must be hand picked. In times long past this system was widely used and absolutely marvelous wines could come from it. In 1972, '73 and '74 many vineyards in Napa were of this system. With the introduction of the California high wire trellis system with its two and three wire models the shoots were forced upright and *then* allowed to flow over the wire and down. Thus, the fruit was now grown in a dark tunnel below layers of foliage – no sunlight, no air and basal leaves shaded into senescence – a totally different environment.

PLATE 10: 2/08

Here I show you a pruned version of the preceding vine. Look at the common, healthy diameter of each spur. Notice that on the central vertical branch that the pruner has smartly left a renewal spur down low. The quality of the spur wood is superior. I have presented these for the benefit of those who have not had the opportunity to view this system and aid in their understanding of my arguments in the text.

PLATE 11: 2/08

Now we are going into areas that should give owners heartburn and farm management operators nightmares. In my wanderings in February, 2008, looking at vineyards I was appalled at the extent of poor pruning, poor farming and a general lack of understanding of the grape vine at this point in time.

At the same time I also saw widespread application of the new viticulture principals with errors only in the details that could be easily corrected with greater understanding and diligence in supervision. There is much to be pleased with – far more than in times long past.

While a few of the shots were at extremes to clearly make a point others display widespread situations. Picture 11 displays shoot length situations that I observed over widespread areas. These vines were allowed to carry crop the preceding year and were harvested. One can see the diameter at the base when growing from stored energy and then later the crop load pulled the cane to a screeching halt. These vineyards were simply not farmed correctly. Vines in this situation are not given enough fertilization, water or missing micronutrients or all of the above. They certainly should not have been allowed to carry whatever crop they did. There is no long-term commitment to vineyard health or quality of fruit. Sometimes in these situations there can be reasons beyond the apparent. It very well may be that owners are in cash flow difficulties and are failing to provide funds to the management company – to the detriment of the farm manager's reputation. It may be that an ignorant owner may be simply setting a fixed unrealistic dollar amount per acre for operations.

The preponderance of this syndrome, however, was on vineyards owned by larger winery corporations or absentee investment vehicle holdings. I think stockholders or shareholders would be shocked and dismayed to discover the dismal level of knowledge or concern on the part of their trusted "managers" and executive. The *lost potential* income from such a cavalier attitude and approach is nearly all "bottom line". So many vineyard costs are "fixed" such that the variable cost to farm correctly is really quite small while the increased returns are mostly to the black bottom line. It is incredible to me that folks who live and die by the up tick of stock or ROI haven't figured this out yet. Of course, short-term mentality rules and future debilitation of the vineyard will be on someone else's watch. Folks, lousy farming leads to lousy fruit and that leads to lousy wine. Lousy wine leads to an eventual situation that Rolaids won't cure.

PLATE 12: 2/08

This is a different vineyard owned and run by a different large winery operation. Again, notice the cane length on these fairly young vines. The spur sites are already beginning to get out of control although it is not uncorrectable at this stage. Notice also the large diameter spur sites in the yokes of the vines. The yoke area of about 12 inches of the cordon wire is unused for fruit. It is repeated at all the vines as you can see in the background. If proper vigor had been generated sufficient cane length would have been available to come back on the second wire to cover the yoke gap as well as another to cover the gap between cordon arms at their ends. Let me put it another way. Those vines appear to be about 6 feet apart. There appears to be about 3 inches between cordon arms and 12 inches or so at the yoke so a wasted 15 inches is associated with each vine. Six feet is 72 inches so 15 divided by 72 is about 21% reduction in potential crop from this situation. Notice all the associated costs are constants. The only variable is a 21% increase in crop just from proper trellis usage.

PLATE 13: 2/08

Again, consider shoot length in the old-style vineyard and the fruit there from. However, there have been good manager efforts to control the spurs on these old vines. They obviously need more fertility and maybe water.

PLATE 14: 2/08

These vines demonstrate how spur site control can be maintained. The vines are about 18 years old. Notice the spur locations and closeness to the cordon. The spring growth will be in proper relation to the second wire. In a picayunish way I find disfavor in sourcing a cane from the 1st spur site. Notice the color and diameter of the spurs which reflect proper growth and light the preceding season. The Syrah vine has relatively larger clusters so this format will generate sufficient quality fruit. The last spur of one vine is at the proper spacing from the first spur of the second. You might note that the cordon arm is tied to the top of the cordon wire – not wrapped around it. Wrapping will cause girdling over time. This picture is to help set a "standard" of what to do and how it can be done. The spur site maintenance discussed earlier will result in this long-term control.

PLATE 15: 2/08

This picture was taken in February, 2008 and displays long-term failure to control spur sites. This condition was far more prevalent in the past than today. Some nearby vineyards in the same condition have smartly gone through the site refurbishment process and are in good shape. The arrow points to a very nice repair site and the large spur should be cut three quarters of an inch above it. The next old spur to the left could be spurred low, a cane from above tied to the 2nd wire for crop next season then removed entirely the next pruning season. The next to the left could be spurred low and the rest removed. Rejuvenation of this vine could be accomplished in two pruning seasons with no loss of crop. All that is required is understanding and diligence in supervision. The vine will actually thank one by giving better fruit in the process.

PLATE 16: 2/08

Same vineyard, same time. The arrow points to the location of the 2nd wire. As you can see the trellis is nearly non-functional in its initial intent at time of installation. If I were to rebuild this vine I would cut out any material in the yoke and cut those big spurs about three quarters of an inch above the cordon. Look directly below the point of the arrow. I would cut the cordon and discard. There is a very nice cane to lay down on the wire to be the new cordon. I would accept the very small crop loss (if any) for the one season. Look at the diameters of the canes further out the old cordon and notice the 12 – 14 inch gap that is providing no crop. The new cordon cane will put this vine in a better productive mode. In the following years this vine could produce nearly double its current rate.

PLATE 17: 2/08

Not the same vineyard but also February 2008. This is a much more common error widely prevalent caused by failure to understand first spur maintenance and to *never* allow growth in the curve or yoke. In picture 17 the first spur is trying to give one a nice little replacement spur down low and should be retained. The rest of the spur should be removed. The next old spur has a good low spur choice and a healthy cane for tying to the second wire. The next old spur can be removed at the three quarter inch above the cordon.

PLATE 18: 2/08

Picture 18 presents a bigger problem. This vine is wasting capital assets and its short term reclamation is not worth the time and effort. I would simply cut the trunk about 8 inches or so below the cordon wire and properly retrain new cordon arms the next season. One loses the crop for a year but it's not much to begin with. Look at the 24 inch gap on the right. That bowed shoulder on the left will always be a problem. Start over and do it right this time.

PLATE 19: 2/08

This picture presents a very widespread problem particularly in larger corporate operations and farm management companies where farm labor contractors are used, turned loose in the field with their own supervisors and the company man goes and has coffee or whatever it is they do instead of supervising. In very young vines the first spur site pushes the biggest and longest. Usually it is the only site that has what appears to be a keepable cane and so the pruner keeps it. The farm manager, either through misguided greed, ignorance or effort to impress the bosses with early yield, is in favor of that. That is the start of the less than optimum spiral. Owners and corporate executives – go into your fields and look – you will see it all over the place. Notice here the 24" gap in fruiting zone after that first spur. The arrow points to one tiny renewal spur then there is another 14 inch gap. Of the 72 inches between vines 38 inches is none yielding! 53%!

This vineyard is owned by a well-known entity, well-funded and heavily staffed with so-called "experts" or "professionals" that have no clue what is going on in their fields – or, if they do, are derelict in their duty. Strong words but warranted. Owners, if you like producing at the 50% level of potential on your capital investment ignore everything I have said. However, if you have anything more than a mild affection for profit or return on investment you might listen. It could be worth your while.

PLATE 20: 2/08

Again, a common sight. This syndrome is caused by poor supervision at the time of training. Notice the dominance of the spur at the top of the bump and the tiny, tiny spurs further out the arm. This should never be allowed to happen. If this condition were to occur in a vineyard one wished to keep then the solution is to cut the trunk and train new cordon arms the next season properly placed. This is a completely non-economic vine. Period.

PLATE 21: 2/08

Again, the new viticulture put into practice with failure to understand the details and the goals. The common misapplication of principals by using the first spur site from the beginning! Look at the diameter of that first site which has continuously sapped off flow to the remainder of the cordon and thus always giving the most appealing cane to the untrained eye. Look at the reduced size of the spurs further out the cordon. A further error is that if a gap exists (arrow 2) any cane on the second wire should go over the *gap* – not over the other spurs.

Another error in maximizing returns is that the cordon arm of the vine on the right should extend all the way to the first spur site on the left vine. That appears to be over 24 inches of foregone production. At 5 feet between vines that is 40% forgone potential income at no particular increase in cost! If there was not sufficient vigor to fill that space then the vines were too far apart and vigor inducing farming procedures need to be utilized. And again, anyone interested in 40% of our factory sitting idle? Or 40% of your capital investment generating no income?

PLATE 22: 2/08

Here is an example of someone getting it almost right. The small errors here are the bowing up then down of the right hand vine at the shoulder and the spur site in the shoulder. Long-term this vine will become a problem child. The left hand vine is in good condition but those two spur sites in the curve should be removed – now. Had they not been there this vine would have generated sufficient cane length to hold some cane on the second wire.

PLATE 23: 2/08

PLATE 24: 2/08

These two pictures demonstrate the effects of farm labor crews allowed to operate unsupervised. Often piece rate payment results in sloppy high-speed work. These pictures show the slashing to two bud spurs with no spur site renewal. Look at the gaps of non-production. This vineyard is only about 15 years, or so, old and is displaying all the progressive signs of the downward spiral.

Notice the ground cover drop in the background. This was drilled in. Someone along the way said winter cover crops were good so cover crop is the way to go and it was done – regularly. However, this vineyard is on rock, sand and gravel – little clay material. There are no sprinklers – drip only. Thus, the only moisture in the main soil mass is rainfall – there is no way to recharge it at will. Cover crop removes the winter moisture. Guess what – a summertime problem and a severe shortening of the life span of the vineyard (see Irrigation Section). I realize that in modern corporate life it is not popular to consider long term but I do point out that incorrect procedures can result in a vineyard that is so ugly that if an entity decides to sell the vineyard rather than tear out and redevelop the selling price could reflect the prevailing bare land value *minus* the cost of tear out! We have seen that occur before at low phases of the cycle. At, say, $50,000 per acre ex land to develop a modern vineyard (some point in the future) it seems to this country boy that preservation of such a capital asset should be on the radar screen. Properly maintained vineyard should have at least forty years of economic life (absent disease). Mis-farmed vineyards can progressively decline to below economic levels in twenty years or less! Errors are *not* one shot deals – they are *progressive* and take their toll over time.

PLATE 25: 2/08

This is a nice example of what the vine should look like with easily correctable errors. The spur sites in the curve should (must) be removed. One spur mid-way on the cordon should be cut back. The canes are good, have good color, and good internode length. A cane should be placed on the second wire to absorb energy with more fruit on this growing vine.

PLATE 26: 2/08

These vines are starting to get out of control. Of course, the spurs in the curves should be removed. Consider the diameters of the spurs at the end of each cordon compared to those at the first spurs. Both vines have spurs that should have been cut back for spur control. These are examples of the common two-bud spur pruning – fast and inexpensive. Tell me – how many fruit clusters are within those BASAL buds? I know of one large management company that knows of the procedure but doesn't "do it because it takes too much time and we're not budgeted for the other work"! Merry Christmas, owners!

PLATE 27: 2/08

PLATE 28: 2/08

This is another system used by some as a modification of the new viticulture. It does take advantage of the more fruitful cane buds as a matter of count and it does give excellent fruit distribution. It provides far more buds that a spur pruned full cordon and the bowing provides more push incentive to the mid-cane buds. There are factors that I do not care for in this format. The spur site selections are extremely limited and thus their repetitive use leads to dominance. I like a longer cordon with at least 8 spur sites such that a two cane is a four year rotation. You can see that the moveable wires are down (they come down at an angle from the right) and ready for raising up the coming season. Given their desired design it does not get better than this in application and attention to detail.

PLATE 29: 2/08

I just throw these three in to show bizarreness, at least to me. I'm sure the owner has some rationale and perhaps it works for him and his customers. The vineyard is quite old. To me it was, at first glance, like being in a strange garden on a different planet – like Australia! They actually looked like some strange coral or creature living at depth. I have never seen the "bush" system of Australia but this must have some basis in it. The system involves tight mechanical pruning and then some quick hand touch-up.

PLATE 30: 2/08

PLATE 31: 2/08

PLATE 32

This picture is of the Ventana pruned and tied and unpruned. In this parcel the cordon arm doesn't quite reach the next vine. That gap provides the tie-down location for the bowed canes. The vines are on their own roots and are 3.5 feet apart. The vertical plane is fully utilized and fruit distribution along the trellis as will be shown in a later picture.

Regard the length and color of these Chardonnay rows on the right. This picture was taken many years ago – I don't recall exactly when but at least ten years back.

PLATE 33: 2/08

This is a different variety of larger size clusters – Syrah. Here the cordon is continuous and less cane is desired. This variety produces well from basal buds and does need fruit thinning during the season – at veraison. Some years it may require an early pass then one late or after veraison. Under cropping or too-early thinning can result in big berries reducing skin to juice ratio.

Notice that the stakes have a coating to reduce corrosion thus lengthening their life.

PLATE 34

Before and after pruning and tying about 1995 on the Ventana. The old wooden stakes would soon be removed. One sees the old A-4 cross arms – no longer used. The use of the second wire is clear.

PLATE 35: 3/08

This shows what we are after – uniform bud push.

PLATE 36: 3/08

Again, we see uniform bud push along the cane and the spurs below. This is an important step in achieving uniformity in fruit at harvest time. See the text for discussion.

PLATE 37

This was taken long ago – see the wooden stakes – I believe around 1981. See the even push along the cordon. This is the 1978 planting of Chardonnay on the Ventana.

PLATE 38

This also was the early days of the new viticulture on the Ventana. The evenness was not what we wished. Notice some buds not pushing on the canes which, of course, leads one into contemplating the energy/bud line of thought discussed in the text.

PLATE 39

The leaves have been pulled on the Sauvignon Blanc on the south side only. The golden clusters are about to be harvested. Notice the greenness and soundness of the upper leaves. The canes are lignified nicely. The fruit distribution is very good and no botrytis. This is the Ventana some years ago.

PLATE 40

Here we are looking at Chardonnay a long time ago – see the wooden stake. This is a lovely distribution of fruit in both the vertical and longitudinal. The leaves were stripped on the morning side to enable light, air and spray thorough penetration.

PLATE 41

This picture displays the Tempranillo grape on the Ventana. The tube encloses a replacement vine that a gopher damaged severely. The downwind, southeast side leaves have been pulled. The gentle morning sun has aided color while the un-pulled afternoon side protects. The variety, though relatively new to Americans, will become very important in the future. The fairly large berried clusters yield well.

PLATE 42

This picture shows Syrah fruit distribution in the trellis with leaf stripping on the morning side only. One can see the use of the vertical plane for fruit separation. Ventana Vineyard.

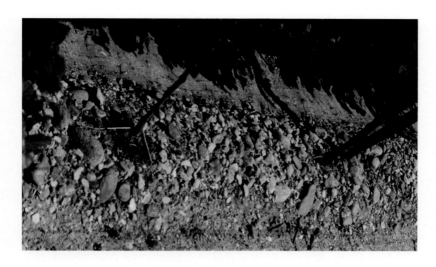

PLATE 43

The natural rocky ground of Ventana.

PLATE 44

This is the BOUZA vineyard in Uruguay. The vineyard is a textbook example of the new viticulture – which has taken over South America. The rocks under the vine row were all hauled in. You can see the clayish native soil at the edge of the rocks and within the cover crop. This block had just been harvested the preceding day. Notice the complete absence of growing tips on the vines and the green leaves for carbohydrate past harvest generation and storage. Tannat is the premier grape in Uruguay – their rightful pride and joy.

PLATE 45

This picture was taken on October 6, 1995 three *WEEKS* after harvest. It, of course, had received an irrigation. This was Chardonnay on the Ventana. The 6 foot rows lift the wind above the vineyard sufficiently that one can see the foliage having grown into the left side of the rows. The carbohydrate buildup and storage progresses nicely for the next year's crop. Once dormancy occurs cane cutters will shear the vines back as "pre-pruning" and sufficient usable canes remain. The vineyard was handpicked.

PLATE 46

This picture was also taken on October 6, 1995 three *DAYS* after machine harvest. With the twelve foot rows (see text) the foliage is rolled to the downwind side and did not grow well into the wind blast. Notice the declining leaves already deteriorating. The ones that look green are damaged and will fall in a few more days. This vineyard was adjacent to the Ventana, was in the old style and a "beater" harvester was used (see the text for the discussion of this).

PLATE 47-A

At Bouza winery and vineyard in Uruguay about an hour north of Montevideo, the Tempranillo vines on the left and Syrah on the right demonstrate textbook perfection. This picture was taken in very early March, 2008 in the middle of their harvest.

Same place, same time. Here I am looking at Tannat grapes in optimum condition. Notice missing leaves in the fruit area which had been removed about 4 weeks earlier.

PLATE 47-B

PLATE 48

Here is one of my simple study tools I use when exploring vineyards. These two pictures were taken in the Ribera del Duero region of Spain during one of my many explorations of Tempranillo. The graph markings make for easy recall later when more sedentary thinking at home occurs. I have learned to be careful about acceptance of assertions on-site in foreign countries because the charm of the ambiance can overpower me.

PLATE 49

PLATE 50

This picture is in Alsace, France with my great friend Gilbert Dontenville. He is a wine grower and these are Riesling vines he is instructing me on. The trees in the background are the beginning of the forest that goes up the Vosge Mountains from which Gilbert, from time to time, takes a boar. His brother, Roland, and his wife, who own the auberge in town, then convert it into the most fantastic meal. To die for! LuAnn and I, and Ken and Kris Gingras have enjoyed Gilbert and Sandri's hospitality many times. Gilbert introduced me into a circle of winegrowers who were professionally interested in exploring new thoughts in viticulture which is highly difficult in rigidly controlled France. We have had many great sessions but I won't bore you with their content.

PLATE 51

These are the Palomino grape vines of Barbadillo near Jeres de la Frontera – the southern sherry country of Spain. Barbadillo is one of the premier producers and their solera is located in the river-side village of Sanlucar de Barrameda – the departure port for the old sailing ships enroute to the new world. The white soil is caliche – a chalk based soil that is very gummy after a rain. Only a few vines can thrive in it and Palomino is one – Pedro Ximinez another. Here you see the free standing goblet pruned vines as discussed earlier. No, I didn't intend to grow sherry in Monterey but I did need to see various hot climate situations in order that my contemplation and theories explained those observations.

PLATE 52

These two pictures were taken long ago in the Pfalz region of Germany near the village of Wachenheim. Notice the bent nail to use as a connector for the trap wire to hold the foliage up. As discussed in the text later on I devised a connector based upon the Christmas tree bulb hangers. We didn't use posts like that and, even if we did, I thought driving the nail through could crack the post over time. I observed those cracks.

The second picture shows their very unique method of lengthening or shortening the wires and of affixing them to the post. Loved it! But I never used it. It really wasn't necessary here.

Earlier in the discussion on growth I mentioned that when the vine was in a luxury condition and growing well the tendrils were longer than the tips. Look at the nail picture and see how the tendrils are longer than the tips.

PLATE 53

PLATE 54

This is in Chile. By spreading my arms I am showing the distance between rows – not trying to fly. Notice no tips are growing. You can see the leaf debris from leaf-pulling of about 4 to 5 weeks earlier. These vines had been harvested a few days before.

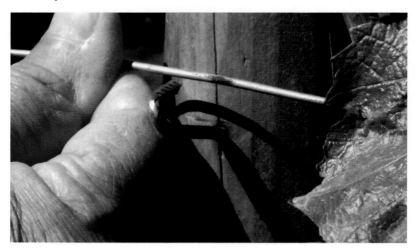

PLATE 55

The bent nail approach of Germany is seen here in Chile. See the split seam.

PLATE 56

This is a scene of leaf stripping and snipping tips – mostly tipping. Look at the row centers behind the workers and you will see the debris on the ground. Notice the height of the vines. This vineyard is just outside the village of Pauillac in the Bordeaux area. This was taken late June. Pauillac is just over the hill in the background. Notice the pebbly soil with lots of chalk rock mixed in.

PLATE 57

This picture is in the area of Margaux. The outskirts of the village are up where the trees are. It is about the same time of year – late June – but it could have been any one of different years. This is on the sloping ground down towards the river from the village. Notice again the pebbly soil and lack of tips. If you look closely you can see some shoots with their tips gone however you can see new ones trying to start. A little later they will make another tipping pass.

PLATE 58

Obviously, this is in the Margaux district. I used my son for height reference. He was seven then and, now at sixteen, he is six feet tall but the vines are still short. My wife, LuAnn, is 5 feet 2 inches so you can get a measurement by comparing her to the vines. Again, no tips and pebbly soil.

PLATE 59

This is a view of Andeluna's vineyards just outside their winery. They are located in Tupungato, Argentina about an hour fifteen south of Mendoza. The views are spectacular with the vineyards on flat desert land backgrounded by the massive Andes soaring above them. The vineyards are immaculate examples of the new viticulture – at least the ones in the area planted since 1999. The wines of Andeluna are stunning particularly one called Passionata. Silvio Alberto is the winegrower – over the growing and winemaking. He is also a professor at the Mendoza enology school. The area is around 3,000 feet above sea level.

PLATE 60

This is the Andeluna vineyard in Argentina. These grapes would be harvested in the next few days. There are a few tips showing but most are gone or pulled to a stop by crop load. The lower leaves around the fruit were removed about four weeks earlier. Notice the pebbles in the soil. The vines are very uniform and meticulously maintained.

PLATE 61

Again, Andeluna in Tupungato, Argentina. The Syrah is beautiful in its bounty and cleanliness. The fruit is presented to the pickers at ergonomic height for ease of picking – in the upper picture. The malbec is likewise in the next. The judicious leaf removal shows in blank areas of the lignified canes. The wines from these vines were magnificent. To my eyes this is how vineyards should look. It is just those of the last ten years that do so.

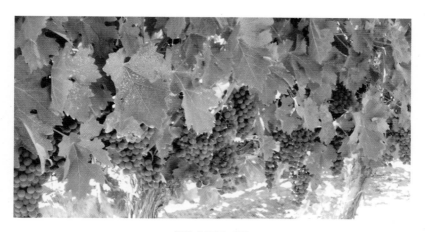

PLATE 62

PLATE 63: 5/08

Here is an example of old-style apples. The old trunks are 40'
apart. The interplantings were done at 20'. See the text.

PLATE 64: 5/08

A follow – on planting system similar to my planting in 1968: 12' x 18'. This was a revolutionary format in 1968. My orchard in this new format received the "Top Grower" trophy for 1977 and 1978.

PLATE 65: 5/08

Another format that came along in the late 70's and 80's. Full dwarf trees utilizing a pole-supported trellis.

PLATE 66: 5/08

Another design of the time emphasizing light penetration. Full dwarf and trellis. Notice rootstock over-growing scion.

PLATE 67: 5/08

These are Labrusca vines near Quincy, Washington – two different but close by properties. The pruning system is certainly unique to my eye but I have no experience with Labrusca.

PLATE 68: 5/08

PLATE 69: 5/08

This is a vine roadside near Lake Chelan, Washington. The "bowing" of the shoulder was somewhat common across the vineyard. This is a classic example. Consider the spur diameters ant their decline. Look at the bud swell at the highest point compared to the others.

PLATE 70

PLATE 71

PLATE 72

Finally – the end of pictures. Nothing more here on viticulture – just some remembrances from about 1976. In the text I spoke of hand crushing, barrels in the barn, etc. That is me against the barn door. I and my guys had borrowed an old basket press for that day. Serafin Guzman, my right hand guy, is crushing Gewurztraminer into the plastic garbage can lined with my Navy nylon mesh laundry bag.

A long time ago.

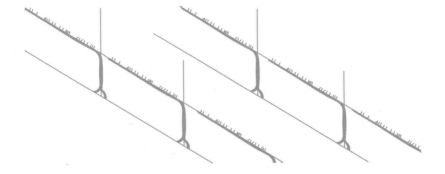

SUBSECTION B

MISCELLANEOUS
WRITINGS

"OWN ROOTS"
by Doug Meador

Prior to about 1867 all European grapevines were grown on their "own roots" – i.e., there was no grafting of varietals to "rootstocks". In that time frame both amateurs and professionals imported "American" grapevines – wine varieties native to North America – for study. Unknown to them, they also brought along in the root mass a little bug native to North America – the Phylloxera. North American vines had evolved in the presence of this rascal and had developed resistance to it. Vinifera – the European vine – had no such resistance! The bug loved the environment, loved the buffet lunch of vinifera and spread like wildfire – most likely from the area of Marseilles in the South. It was devastating.

The French tried everything to save their precious vineyards. Hybridizing – crossing vinifera with American – was tried diligently to no avail. However, some of these "French Hybrids" became of use in the Northeast of the U.S. The French soon discovered that using American roots and grafting vinifera on top provided the only resistance to the bug. Subsequently, hybridizing "rootstocks" was found useful for various purposes and soil types. It was also found that any vinifera in parentage reduced resistance to phylloxera. Most – not all – but most European vineyards were replanted to vines on American rootstocks.

The resulting vines (and wines) were nearly true to the original – not completely but "nearly". This "nearly" business is because the rootstocks interact with their soils differently than pure vinifera varieties on their own roots in a given soil. This is a mechanical thing and affects the associated results. The rootstocks do not cause genetic changes in the top – or "scion" as it's called. The resulting wines are Chardonnay or Cabernet or Pinot or whatever has been grafted on the root.

Many European writers of times past who have tasted current wines of before and after grafting have asserted differences. Many have bemoaned the necessary transition, asserting loss of certain subtleties and complexities associated with a given terroir – in essence, a loss to some degree (apparent to them) of distinctiveness of terroir.

Let me give an example on the technical side how this will occur. A wine is the integrated result of a 'grape' and the winemaking techniques. Lets hold the winemaking procedures constant (historically normal) and discuss the "grape". In a given terroir (which is everything about location – not just the "soil") a self-rooted vine will interact in a fashion unique to itself. The associated grape will reflect its foliage-to-fruit ratio, vine vigor, its ability to extract (or not) micro nutrients (if present), minerals (if present), moisture, etc., etc., etc.

If we now insert into the equation a rootstock under the vines we have changed the vine/fruit relationship. By definition – a rootstock is different than the variety on top. Thus, the entire

Monterey–Arroyo Seco–Estate Wines
Ventana Vineyards • 2999 Monterey-Salinas Hwy #10 • Monterey, Ca 93940
Telephone: 831-372-7415 • Fax: 831-655-1855
www.ventanawines.com • e-mail: info@ventanawines.com

Apr 23 2007 2:20   P.01

relationship of the plant to the soil is different than if the plant were on its own roots. Notice that there is no assertion of necessarily "better" or "worse" – just different. It very well may be that in a very weak dry area a moisture scrounging rootstock could improve a grape in a dry year with no irrigation!

Another example would be the utilization of a rootstock that has a definite difficulty in gathering zinc leading to small leaves and set problems on the top. One major rootstock has - we now know – exactly such a problem. Absent corrective additions of zinc this rootstock would create substantially different fruit.

Most rootstocks are of native American varieties (or hybrids thereof). However, even with vinifera roots – but different variety – I have seen differences in the appearance of the top. One example is Chardonnay grafted onto Zinfandel roots. The Chardonnay clusters appear bigger and longer – more in the shape nature of zin! Yet – they are definitely Chardonnay in every other way – apparently.

At Ventana, for a variety of technical reasons with which I will not bore you, most of our vines are in the ancient way – on their own roots! We have been conducting rootstock exploratory trials since 1974 and do not yet have definitive answers. So far, our observations are that by far the best and most unique quality comes from vines on their own roots. We have also learned that we must match certain varieties to certain soil characteristics – a process that we have persued for a very long time. For example, some varieties have very high natural vigor and these we plant on soil with the most rock and least nutritional aspects. Thus, the vigor is restrained, the berries are smaller and, therefore the skin-to-juice- ratio is higher. As red wines acquire their "essence" and "extracts" from the skins, smaller berries yield wines of more depth.

Conversely, we like to grow varieties of less natural plant vigor on soils of more nutritional character – or feed them more. This works toward balancing the foliage and fruit. We also plant these types closer together thus asking less of each individual plant.

All of these viticultural procedural differences allow us to maximize and express the full range of characteristics unique to a given variety – from its roots through its fruit. The resulting wines is a full expression of the given variety within the Ventana terroir.

This pure expression of the varietal is not widely sustainable in the United States or Western Europe today. The phylloxera bug has changed all that. The Ventana Vineyard is one of the very few locations where the old-world pre-phylloxera vinifera grapevine thrives. In some support of the "own rooted" merits, might be the observation that Ventana Vineyards is now more than twenty consecutive years of gold and silver medals on its Chardonnay and Riesling grapes. Other varieties also have long strings of awards. This would tend to indicate merit in the views of those observers of long ago – something was lost with grafting.

It is the "location" – or terroir – of the Ventana that allows us to bring to you the pure experience of each of our varietals. Enjoy!

Doug Meador

"ON TERROIR"
AN OPINION

BY DOUG MEADOR,

The French word "terroir" has become a current "buzz" word among the English speaking wine community. Of particular interest to me is its spreading use among American winegrowers. Of first concern is the meaning of the word, of second concern is its applicability in any real sense to the American situation. Before we can address any definitive discussion on the merits of the concept it seems to me that we must tightly explore the word and the context within which this concept evolved. "Terroir"- rational or mysticism? The explanation must begin at its origin-which is within the historical french experience.

Let me first present a few current examples of definitions as presented by James E. Wilson in "Terroir":

A) **James E. Wilson:** "Terroir" (page 55) (Mr. Wilson is a geologist)
"The true concept is not easily grasped but includes physical elements of the vineyard habitat – the vine, subsoil, siting, drainage, and microclimate". He goes on to assert a "spiritual aspect." He precedes the definition with "...lighthearted use disregards reverence for the land which is a critical, invisible element of the term."
B) **Matt Kramer:** He refers to a "mental aspect" of the term – in addition to all physical attributes –"...winegrowers feel each terroir should be allowed to be itself and produce the wine for which nature endowed it." In addition it is asserted "The winemaker's vinification style is permissible so long as it does not substitute for terroir!"
"That is, vinification should not make the wine taste significantly different than the 'natural' wine that would be produced from a particular tract."
C) **Hugh Johnson:** "the land itself chooses the crop that suits it best."
D) **Gerard Seguin (Bordeaux enologist):** By his definition quality terroirs are where the habitat permits complete but slow maturation of the grapes.
E) **Daniel Querre, St.-Emilion grower:** He questions any attempt to explain a particular terroir if only its obvious physical conditions are described. He resorts to "something precious – unknown".
F) **LAROUSSE's "Wines and Vineyards of France"** – "it being the link visualized by a consumer between his wine and the winegrower who produced it.
G) **The Economist:** (British) describes how the French use terroir to counter efforts by the E.U. to deal with wine as a "brand". Here the French argue severely limited environments as the terroirs.
H) **Seguin-Moreau:** In discussing sources of oak uses the phrase "Terroir d'Origine". Here they are referring to flavor of the oak suitable for wine barrels.

THE FRENCH FOUNDATION

In the historical scheme of things in France we can observe many important fundamentals and generalities important to our understanding of the term.

A) ISOLATION
In general, the various french major wine regions are separate, the one from the other. Within those regions often districts and communes separate themselves one from another.
Even today – as well as more pronounced historically – there are commonalities within communes with respect to viticulture. Within a district the functional farming procedures are essentially identical on respective varieties. In fact, in most areas these have been codified – no irrigation, x square feet per vine, variety allowed, pruning method, time of harvest, quantities allowed, etc., etc..
However, as we move from district to district these rules and procedures will change – even within the same varietal panoply.

1

This is to be expected. As one farmer determines something "good" neighboring farmers observe and – after initial jealousies – copy. You can't hide a farm. Over long term one would expect within a small area the procedural deviations to be quite small. The French have simply gone one step further by making it "law" or "regulation".

As with viticulture, winemaking also tends to follow the same procedures within a district or commune. The yeasts were natural to the area, the economics similar, the equipment and facilities generally the same (with, at times, an important difference addressed later) and artistic tastes culturally uniform. Again, the rules and regulations apply uniformly within the district – no acidification, no chaptalizing, one not both, time in barrel, time in bottle, etc., etc..

A further consideration is plant material makeup. Until the late 1800's all french vines were on their own roots. Vineyards were planted or replanted using vine material locally grown – often "layered" over from an adjacent vine. At least, it was taken from plants nearby elsewhere on the property. Thus, the vines were locally adjusted and, over time, often clonally (a word not used in history and a concept not recognized) unique – within a colony.

B) SOIL

Seldom does mother earth give us a uniform soil mass over any reasonable area. There are always differences even one step apart. The "aspect" – angle and orientation to the sun varies. The composition varies. The incidence of rainfall varies.

The nutritional level varies. Etc., etc..

For simplification:

Let me now tender an equation:
$$C + S + P + FP + WP = W$$
Wherein: C = Climate
S = Soil
P = Plant
FP = Farming Procedures
WP = Winemaking Procedures
W = Wine

If we consider, historically, as constants (in the reasonable run) general S, P, FP, and WP then the only variable is C (climate). Knowing nothing other than math (not wine) one would project that as C varied into or out of optimum with the others there would be "W" that was good, bad or indifferent – that is, good years, mediocre years, and bad years. And that is exactly what we see in life – look at all the published "Vintage Charts".

But there is more here. On a micro-basis S is a variable. Not that it perceptibly changes (absent human intervention) yearly but it is a constant differential from adjacent sites. Thus, at different C's the inter-relationship with the constant P, FP and WP will result in different W.

Now – some points to consider given the above. If the general rain pattern in an area governed by the "no irrigation" rule (in history it was governed by no technology to accomplish it) is – say – every two weeks then micro-differences will become important. For a given variety on its own roots those soil types which will hold and supply adequate water for at least two weeks will be construed by farmers as "good for grapes". Those that cannot, will not. We begin to see some rational behind "terroir" – differences according to site.

Let's consider chemical composition. For example, what if one micro-location is slightly deficient in zinc compared to others nearly. One effect of this can be small berries. As a matter of physics the smaller the berry the greater the skin-to-juice ratio. Thus, wines from these grapes would be higher in extracts and color. If those factors were valued by humans, that spot would acquire some cache over time. Again – a rational basis for terroir.

C) TIME

Many of these differences attributed to terroir are so miniscule that a substantial amount of TIME on specific human beings must pass in order for the recognition of certain characteristics as a function of site to occur. Once recognized others can be "taught" but that too takes time and experience. The French have had the time.

However, many "recognized" flavors COULD come from other factors. Consider the genetic instability of Pinot Noir combined with historical planting methods. Could the differences between Fixin and Gevrey be due to a profoundly different polyculture of Pinot Noir? Or to slightly different farming procedures? Or to a slightly different climate (micro-climate)? These are nowhere addressed.

Age of plant material is also a factor in flavour generation. There are substantial differences between young vines and old. Which is definitive of the "terroir"?

D) WINEMAKING

A flora natural to a given area will evolve a complex suitable to that environment. In using native yeasts certain flavours will occur. The flora from the wineries and their debris will reinforce that prevailing colony. Also, however, certain strains can be housed within wineries and fermentation infected at the time of entry to the winery. Monopoles could be questioned in this instance while poly-producer vineyards not likely.

In general, however, winemaking within a commune was historically a constant. The deviation may be in that long ago specific properties became tres chic and as a result more profitable. Those sites were able to change some things from their neighbors – lower profitable crop, newer barrels, better equipment & facilities, more detailed hands-on winemaking, less sugar, more sugar, better pesticides, etc., etc.. The same effect can be seen as a function of owner – richman's toy or working farmer. But – in general – the procedures were/are uniform. Today – it is still the case within regions. Across regions it is becoming even more the case as schools essentially all teach the same thing.

Thus, given as constant WP, there would be a definite rational for site specificity on differences as "terroir" - good, bad or indifferent but all within the specific WP. Change the WP and those differences may change.

In summary of the French experience, I submit the idea that there is rational (right or wrong) for the concept of terroir in the French mind. I have not expounded on all the parameters involved in the analysis but you can certainly see the foundational raison d'etre of the concept as it evolved.

Let us not forget to keep in mind elements of "marketing" motivations during the discussion. IF there is a desirable "terroir" in a site and you own the site – you have a monopoly!

However, in modern times I am not so sure about complete justification for the concept. In or around 1863 – phylloxera was introduced to Europe. Subsequent grafting to rootstocks in defense changed the historical relationship between plant and environment. Modern fertilizers change the soil and offset nutritional balances. Micro-nutrients further alter the innate relationship. I will address these and more in the next section.

THE MODERN COMPLEX

Now we come to modern viticulture and winemaking – the emergence of the new world, changes in the old world, modern technology, new procedures, new artistic tastes, modern finance, etc. – and we have some difficulties trying to apply an ancient concept to our era.

The concept of terroir evolved in a time of long term stability (technology of agriculture) or stagnation – your choice of term. It was surrounded by de facto constants. The practitioners were life-long, generations-long people of the soil, focused for eternity on their part and area of the earth – and only that. Many never traveled 20 miles from their place of birth. The earth was their existence, wine their soul.

Today, in France, since phylloxera, most (not all) vineyards are on rootstocks. This necessary condition totally changes the interaction of the vine with its site. The historical mystical aspect of "nature choosing the crop that suits it best" or natural selection by region or any reference to empirical history is meaningless. Now the considered relationship has to include the specific rootstock and additionally its relationship with the scion.

"Which" rootstock is also important as each has different characteristics. What, for example, would be the effect on grapes from Le Montrachet if a high-vigor rootstock were used thus changing the foliage to fruit ratio? Can that be compensated for by intervention with trimming and leaf-pulling? What about a poor zinc-gathering-rootstock on a site where no zinc deficiency previously existed?

3

We hear regularly – with no surprise – of vineyards (of a given variety) planted with one rootstock on high ground, mid-slope another, and bottom another. This is a
de facto recognition of change in terroir AND A DELIBERATE EFFORT TO CANCEL IT!

All we have left of this P segment is the varietal to climate relationship.

Let us now introduce irrigation. This modern facet eliminates – where allowed – the entire segment of historical relationship of wine to soil water-supplying capabilities. Of course – it also eliminates crop failure from drought. In fact, it also allows greater crop size – though this is possible without altering the foliage to fruit ratio. Much of the New World could not grow grapes without irrigation. Now – does a whole new "terroir" component arise within the irrigation model? Probably. Over time and only within operations that consistently avoid water mismanagement (whatever that is) the differences may emerge – if anyone is looking.

"Over time" leads one to observe that the modern corporate farming approach does not lend itself well to long term observation and experience on small sites. How does one go about observing the wine over time from small sites on these operations? And "why" observe if the field procedures change constantly?

New varieties in new areas, new clones in new and old areas, monocultures versus polycultures- these all play a part in our new ways. A clone that does well in Dijon finds itself a current "darling" and in three years California has 25,000 acres of it planted on every rootstock imaginable (except AXR) and from Mt. Shasta to Death Valley- all of which remind some industrialist of "soil just like I saw in Burgundy" usually after "years of searching California for just the right soil" (about 5 minutes from their house of forty years).

The amazing thing is that our current knowledge allows a much wider growing area for Vinifera than mother ever intended. Fungicides allow eastern U.S. growing. Certain techniques allow winter cold areas to have less impact. Mildew is controllable – usually, Botrytis succumbs. Rootstocks allow growing on caliche, irrigation allows desert growing or growing in a rockpile.

Please note that these, and others, are all "human intervention" actions. Today, a zinc deficiency gets zinc applications, Boron Boron, etc., etc.. How can there be "terroir" in the full (including mystical) French historical sense when every nutritional condition can be – and must be – corrected? Modern financial concerns dictate that – except for the toy farms.

The old propagation techniques led to locally adapted polycultural plantations- most apparent in the Pinot complex because of its notorious genetic instability – genetic drift or mutation (which is a form of environmental adaptation/selection). However, I think the adaptation – over time – is done by the other varieties though less dramatically.

We have mono-cultures – one size fits all soils. Rootstocks under them are poorly understood by the best of us. Given that, vine spacing and vineyard design is shot-gunned in at best. The desire/need to "fill the trellis" under these conditions leads to every level from severe over-cropping to severe under-cropping. These conditions destroy any concept of terroir.
We do see some regional generalities emerging. These are mostly explainable by weather/ climate patterns as far as I can tell. Napa and Sonoma – in some areas – have been at this for some 150 years though a vineyard of today is nothing like those of then. Yet, some areas have developed some repute. Whether, blind, experienced judges could pick out a "terroir" characteristic – I'm doubtful. There are many very old-vine vineyards in California. It would be interesting to have a blind evaluation of wines from those – say Zinfandels or such – and have experienced judges judge for "terroir". I suspect the quality levels are a function of "old vine" and vineyard design rather than specific terroir identifiable.

Then we come to modern winemaking – particularly the "international style" and the "California style" – whatever those are!

One main component is the love affair with new oak. Reread Matt Kramer's definition – vinification should not detract from the natural wine produced by a tract! To express a "terroir" you cannot

4

turn the wine into a liquid toothpick! The loading of oak covers any possible attribute of terroir. I have heard folks speak of terroir then tasted the wine. Many, to me, are wines only a termite could love – and some judges.

The other procedures of modern winemaking give us many safeguards – and those are necessary in today's financial world. But, sadly, everytime we take something out of a wine nature requires the removal of others along with it. The more "market safe" we make it the more stripped of some characteristics it will be. Tartaric crystals are one example, fine sediment another. The question is – to what extent do these procedures emasculate whatever "terroir" aspect is extant?

We know that different cultured yeasts result in the formation of different aromas in wines, different textures on the palate and even different color intensities. Some wineries use a heat process for color and essence extraction while others use bleeding and long macerations or pre-fermentation "cold soaking". Why? To change the characteristics of the wine in ways – to them – that are significant – else, why the effort?

In summary, I think the French historical evolution of the concept of "Terroir" has merit and has a rational basis even if parts of it cannot be yet explained. The mystical aspect is simply a declaration of human lack of knowledge – so far. Someday we will be able to explain the "unknowable something".I'm sure. Within their framework of essential constants the differences in small areas due to physical attributes are sure to arise and become apparent to long-term stewards of the sites.

I specifically reject the mystical or "something" unknown aspect as not worthwhile to analyze because it is not definable. As something not definable it is not subject to rational thought. Further, the objective parameters have not been explored thoroughly by humankind. These parameters have great range of variables and could easily incorporate the "unknown" once investigated.

In modern times and in America and Australia – in general – I am not a strong proponent of the concept. There ARE particulars where micro-site "terroir" seems to be emerging. Our own vineyard (single unit) is larger than many French communes. With essentially the same cultural practices we are seeing some consistent remarkable differences in very small locations within – after twenty-eight years. However, we also see substantial differences as a function of vine age.

We do observe larger area commonality due – I think – to climatical effects – not soil as the soils vary within the subject areas.

We can observe some locational aspects when winemaking effects are neutralized to some degree. Andre used older American barrels at BV for decades. Today, the Georges de Latour receives two-thirds new French oak. But the old BV shows a continuing theme – of the soil or site.

Modern viticulture & enology – here and in the old world - are in such a state of rapid change (i.e., a lack of constants), people are so new or are transient, new lands involved, and new plant materials so pervasive that to a very large degree we are profoundly premature in utilizing the full historical concept of terroir in any but the most basic sense of physical attributes of a site or region.

Thus:
a) Definition: the physical attributes of a site or region
b) Terroir as an historical concept had a rational basis based upon constants
c) The mystical aspect was a sign of human knowledge deficiency.
d) Terroir in the modern world should be used sparingly (this includes France).
e) There are very small sites of long standing in California beginning to emerge – but very few.

Ventana Vineyards Winery
2999 Monterey-Salinas Highway #10
Monterey, CA 93940
(831)372-7415, Fax- (831)655-1855, Web Site: www.ventanawines.com
E-mail: info@ventanawines.com
5

Sauvignon Blanc: "a marvelous creation"

DOUG MEADOR

The
Leading
EDGE

The Sauvignon Blanc (S.B.) grape is a marvelous creation, in my opinion. This grape is the foundation material for exquisite white wines of Bordeaux, Poitou, Haut Poitou, Sancerre, Pouilly Sur Loire, Menetou — Salon and California.

To my palate the S.B. based wine is a chameleon, taking on characteristics from the food with which it is enjoyed. It is far more compatible with foods than its more famous cousin, the Chardonnay — particularly so when the Chardonnay is done in the fat, overblown style. Where the Chardonnay demands stage center, the more gregarious S.B. looks for partners to blend with, both in the process of becoming wine and in the process of approaching humans. It wants companions, it wants association, it is the Tabernacle Choir to Chardonnay's prima donna, prizing harmony over individuality. It is truly a "dining wine", happiest when enhancing a total cuisine experience, least happy when forced to solo. It possesses an awesome range. With delicate shellfish, it becomes diminutive, focuses the flavors and pushes them forward. At the other end, with hot, spicy foods, it rises to the occasion, quenches the fire, attempts to revitalize taste buds just assaulted by Szechuan peppers or Cajun spices, and bursts forth with flavors only hinted at when in their company. Amazing grape!

Who, and what, is this creature. Today we will explore this rascal, at least as far as I think I understand it — which is, admittedly, imperfectly. As it evolved in France, let's start there, specifically, in Bordeaux. In white Bordeaux, the S.B. is only part of the wine. The other main grape used is Semillon, usually in the range of 40% to 60%. Far more Sémillon is raised there than Sauvignon Blanc — in fact, it is fair to say that Semillon is the primary white grape of Bordeaux, not Sauvignon Blanc. Another grape used in the blend, up to 15%, is the Muscadet.

Despite the spelling this is not a "Muscat" grape, with all that that implies. The "great white wines of Bordeaux range up to about 65% S.B., going by planting ratios of vines in the vineyard. A more common range is 40-50% S.B. As I understand it, the Frenchman's sensibility on the subject is: The S.B. is for the wine's youth, the Sémillon for

structure, and ageability. Further, the Semillon is required to "control the savageness" of the S.B. The French use one word, "Sauvage", that encompasses two meanings in English — namely, wildness and savageness. Correctly, I think, as the grape can be aggressive in its nature if handled callously. Thus, the blend. The seeking of companions to become civilized. In the marriage process, the parts sometimes struggle and go through difficult stages before integrating the one into the other. Viewed during those stages, the facade presented can be unpleasant. Time works wondrous things and the wine can evolve, in the hands of a skilled guide, into a sublime liquid. Supple, responsive, it is eager to participate in the orchestra of the table.

Some 100 plus years ago the Sauvignon was introduced to California. The legend has it that the grapes came from the famed Chateau Yquem, and though first planted at Cresta Blanca it quickly moved to the Wente vineyard nearby. The Sauvignon Blanc of America, as distributed near and far, came from there and is known among wine farmers as the "Wente clone" because of its growing location. The grape did nicely in the Livermore Valley with its stony soil and very warm climate, causing — as the story goes — one Count Lur Saluces (upon tasting the wines in California) to comment how pleased he was to see his children's performance in the New World. A lovely and charming story, in its own right.

But, what does all this mean to us as

producers in the modern age? Sorry, folks, but now it starts getting complicated and problematical, and probably arcane. Because, all was not well in Eden. First of all, Sauvignon Blancs of California do not taste like Sauvignon Blanc of Bordeaux. I'm absolutely NOT making qualitative distinctions, as both places produce lovely white wines. I have been a lover of Sterling Sauvignon Blanc, especially when made by Ric Forman and then by Sergio Traverso, and I have been a lover of LaVille Haut Brion. That goes back a long time. But, as a researcher, I wanted to know why. As an American Winegrower, I wanted to know why. And so the exploration began.

California does not have a history of cold climate viticulture, in spite of what you may have been told. We have a body of empirically — acquired farming knowledge based upon very warm climate experience and almost all of whatever research was done was done in hot climates with minimal rainfall during the growing season. That body of knowledge was acquired through a long period of history when America made essentially three wines — red, white and pink. The procedures of farming and plant material selection were based upon, generally, disease freeness and yield rates. Flavor composition was not a criterion. The American society then put no requirements upon the producer for "varietal character" as it does now and has for the past decade, or so. Those few who knew bought French wines. Cold climate viticulture had its modern advent in the late sixties and early seventies — the time of the great planting boom.

Cold climate viticulture is a whole new ball game. Out at the margin, Mother Nature can grant greatness. She also exacts a very high penalty for errors, lack of knowledge and lack of skill. Sometimes she even penalizes for no discernible reason. Philosophy aside, cold climate viticulture forced research upon us because it magnified errors and problems otherwise blurred in warm climates.

Warm climate viticulture is relatively forgiving. With that thumbnail sketch of a background, let me take you through the thought development concerning S.B. In this cold climate (Monterey County, Calif.) I could get nothing but asparagus juice from the existing clone of Sauvignon Blanc. Period. Brutal, but true. While other varietals succumbed to farming procedural changes this one gave me the intuitive feeling that even though I could change it, I wouldn't be able to change it satisfactorily. The first stage was an in-depth study — to determine "the nature of the beast".

First of all, why Yquem? The main grape of Sauternes is Sémillon, not Sauvignon Blanc! Chateau Yquem uses it mostly for a dry white called simply "Y" (pronounced e-grek), a wine with no particular pedigree except that it is from Yquem. The source was not one of the vineyards famous for its dry whites, such as LaVille or Margaux (Pavillon Blanc). Suspicion number one.

Second, the source of knowledge was from a lady's diary who had written friend Count Lur Saluces for cuttings. Next thing we know, vines are at Cresta Bianca. Documented source? — perhaps. Suspicion number two. Then, the grape underwent a century or more of a process of field selection for propagation and that is, in modern terms, a recipe for "clonal selection". Read all the old papers and books on "How to Select Cuttings" and you will see what I mean. It truly rewards brisk mutations venturing towards disease resistance and high yields — only. Grapes are notorious for genetic drift via bud mutations, and this one is no exception.

So, even if we grant that in antiquity truth, and if we concede that a reasonable clone of Sauvignon Blanc did cross the Atlantic, a strong case can be documented for procedures leading to what I call "Schizophrenic Clonal Selection". What all that nonsense means is that the farmers responded admirably to the criteria the American public set for them. It just happens that consumers have changed their minds and are setting new criteria for us in modern times. Nothing wrong with that — it's called life, or growth, or progress, or maturing or something. It's irrelevant — our job as winegrowers is to respond.

As we explored the S.B. in this country, many anomalies surfaced. First, it has a cropping problem. In many areas it wouldn't set a proper crop — I know of one vineyardist that didn't get a commercial crop until his ninth year — lots of growth

— no crop. The solution was found by many growers by going to fancy trellises. Still, it is, in *general*, a shy bearer in most cooler places. Yet in Bordeaux, the trellises are not fancy — no real differences exist between S.B. and the Cabernet complex when grown on the same vineyard. Yet Bordeaux is cold or cool seven years out of ten. Suspicion number three.

It tastes different than the S.B. of Bordeaux. Tell me true — have you ever had a white Bordeaux that smelled or tasted "grassy", "herbaceous" or of "asparagus" (these are common descriptive phrases of California S.B.)? Now I have found wines of Poitou and Haut Poitou that are labeled Sauvignon Blanc and fit those phrases but I've never had one from Bordeaux, even from the coldest of years. Consider the S.B. of the Loire — "wet wool" and "cat urine" are often-used phrases, but not "driving grassiness" or "asparagus" — never. In Bordeaux, with its cold and its rains, the S.B., at its best, is described as "figs". Why "asparagus" in our cold climates? Suspicion number four.

In all the literature of the Bordelais that I have perused, I have found no reference to a cropping problem with S.B. except one — and that was dealing with an overcropping problem! Not one mention! They write about Cabernet Sauvignon with the same problems we see here. The French have written about everything under the sun — and if the same thing that happens in California, or Chile or Argentina, or Spain. But nothing about difficulty getting a crop of S.B. — nothing, not one word! Suspicion number five.

Enough technical stuff. In summary, my grandfather once told me that "If it has spots like a leopard, growls like a leopard, and has claws like a leopard — it might be a leopard". In this situation, a strong case can be made for the widespread dissemination of a clone of Sauvignon Blanc that does not rise to our modern standards and desires when farmed in cold climates. That's all it means.

Or, this variety, because of my intuitive belief concerning successfully modifying the existing S.B. with farming techniques, the Ventana team embarked on a parallel project to find a clone of S.B. amenable to cold climates. And while our other procedural research projects have been tremendously

enlightening (though leading us into some very esoteric and controversial areas), the "Wente clone" of Sauvignon Blanc refused to knuckle under. Procedural changes in farming techniques have led to Ventana vineyard producing Gold and Silver medal winning Chardonnays and White Rieslings for 10 consecutive years, Golds and Silvers for Petite Sirah, Pinot Noir, Gewurztraminer, Chenin Blanc, Gamay, Pinot Blanc, Cabernet Franc and Cabernet Sauvignon — all from the same single piece of property! Yet, Sauvignon Blanc resisted.

Another clone of Sauvignon Blanc was found that appeared to have merit. The grapes on the vine taste magnificent. The first experimental wine from 25 vines was made in 197? Excitement! More vines grafted over. More wine. Two rows grafted. More wines. Cold years, warm years, wet years — and it was always beautiful. The existing clone planted nearby — always asparagus juice.

Because of our work, other farmers called this grape the "Ventana clone" of Sauvignon Blanc to distinguish it from the Wente. We like that and have adopted that terminology. We have about 14 acres producing and in 1987 tore out all of the Petite Sirah vines to make room for more. We have provided wood to other growers and the state now has approximately 125 acres planted — all in colder areas.

(Meador, a member of a lively panel at the "World of Wines" Festival at the Ritz-Carlton in southern California in November, contributed this insight into a different — and, he is convinced — superior clone of Sauvignon Blanc. He is ambiguous about its origins other than to say it is not a Wente sport. As to its quality, Meador said his 1985 Sauvignon Blanc from the new clone got medals at every American competition, save one in 1986. In 1987 it got gold medals at the San Francisco National and at a worldwide judging of the varietal in New York City last June.) ◼

A RECORD YEAR FOR HANNS KORNELL

Sales figures for 1987 show the Napa champagne house running 65% higher than 1986. Paula Kornell, vice-president/marketing for the family-owned winery, noted four reasons for the best-year-ever sales:

(1) Continual improvement of the quality of the cuvées;

(2) A strong marketing program;

(3) Strong distributor response; and

(4) Strong growth across the board for *méthode champenoise* wines.

Kornell said she expected 1988 to be an even stronger year for champagne sales.

KASMATIS — *Viticulture Extension Agent* — 8/7/73 — *Calif of the time*

VARIETY	VIGOR & HABIT GROWTH	CLUSTER SIZE	PRUNING METHOD	EXPECTED TONNAGE & HARVESTS	
Chardonnay	V. Trailing	5-10 per lb.	Cane	2-5	E
Riesling	M. Upright to spreading	5-6 " "	Cane	2½-4½	M
Chenin Blanc	V. Spreading	3-4 " "	Cordon	4-7	M
Gewurtz.	M. Upright	8-10 " "	Cane	1½-3	E plus
Pinot Noir	W-M Upright to spread				
		7-9 " "	Cane	2-3	E plus
Pinot N. (GB)	M. Upright	6-8 " "	Cane	3-4	E
Zinfandel	M-V Semi-upright	1-5-3 " "	Cordon	3-5	M -
Petite S.	M. Spread	2-3 " "	Cordon	3-5	L plus
Cabernet S.	V. Spread	6-8 " "	Cane	2-4	L
Gamay	M. Upright	2-3 " "	Cordon	3-5	M-L
Merlot	W-M Spread	5-7 " "	Cane	2-3	L plus
Flora	M. Spread	3-5 " "	Cordon	3-4	E -
Sylvaner	M. Semi-upright	4-5 " "	Cordon	3-5	M plus
Pinot Blanc	W-M Semi-upright	4-6 " "	Cordon	2-4	M -
Grey Riesl.	VV Spread	3-5 " "	Cordon	3-5	E plus
Semillon	M Spread	2-4 " "	Cordon	3-5	M plus
Sauv. Blanc	VV Trailing	4-6 " "	Cane	2-6	E -
French	VV Spreading	3-4 " "	Cordon	3-8	E -

JANE - REMEMBER what I SAID ABOUT YIELDS
& QUALITY: HeRe IS A SPANISH CHART - YIELDS
& QUALITY RIOJA - theIR opINIEN. COMPARE
Doug

↓

Año	Hectáreas Producidas			Producción (Kg/uva)	Rendimiento (Kg/Ha)	Elaboración (Litro/vino)
	Tinto	Blanco	Total			
1985	29903	9094	38817	241296770	6319	173346717
1986	29938	8079	39015	173529246	4448	119830258
1987	30206	9066	39271	186151310	4740	133749709
1988	33040	8997	42045	180410559	4291	131082102
1989	33851	8840	42891	223279641	5230	160609524
1990	34132	8689	42851	225636496	5266	161242940
1991	34381	8509	42889	213410823	4976	145345353
1992	35846	8227	44075	214637001	4870	149938412
1993	37528	8247	45775	249738789	5456	173920771
1994	38955	8238	47193	241609232	5121	168843546
1995	39257	8090	47357	303643224	6412	217910958
1996	38378	7923	47301	340408707	6919	244466446
1997	39920	7844	47785	354341289	7528	253574457
1998	40679	7709	48388	380795587	7870	273560471
1999	42522	7484	50006	301150847	8022	216241745

EVOLUCIÓN DE HECTÁREAS PRODUCTIVAS Y

PRODUCCIÓN EN LA D.O. CALIFICADA "RIOJA"

PARTICULARLY
YEARS
'94
'95
'96
& '98

AÑO	CALIDAD	AÑO	CALIDAD	AÑO	CALIDAD	AÑO	CALIDAD
1928	MB	1946	N	1964	E	1982	E
1929	N	1947	MB	1965	M	1983	B
1930	M	1948	E	1966	N	1984	N
1931	MB	1949	MB	1967	N	1985	B
1932	N	1950	N	1968	MB	1986	B
1933	N	1951	N	1969	N	1987	MB
1934	E	1952	E	1970	MB	1988	B
1935	MB	1953	MB	1971	M	1989	B
1936	N	1954	E	1972	M	1990	B
1937	N	1955	E	1973	B	1991	MB
1938	M	1956	B	1974	B	1992	B
1939	N	1957	N	1975	MB	1993	B
1940	N	1958	E	1976	B	1994	E
1941	B	1959	MB	1977	N	1995	E
1942	MB	1960	B	1978	MB	1996	MB
1943	B	1961	B	1979	N	1997	B
1944	B	1962	MB	1980	B	1998	MB
1945	M	1963	N	1981	MB		

E: Excelente

MB = Muy BUENA

B: Buena

N: N

M: Mediana

VENTANA VINEYARDS
"The Most Award Winning Vineyard in America"

August 28, 1989

Mr. Richard Smart
MAFTECH, RUAKURA
AGRICULTURE CENTRE
Private Bag
Hamilton, New Zealand

Dear Dick:

Enclosed please find a copy of an article in "Farmer's and Ranchers" newspaper. It quotes you contending that close row spacing in high vigor areas is not the way to go (I've flagged the line in yellow). More often than not, articles quoting me are grossly in error. Thus, I first ask if this quote is, in fact, your opinion.

If so, let me address that issue with some thoughts of mine that you may wish to consider. If, in your viewpoint, there is flaw I would greatly appreciate having it (them) pointed out to me.

In essence, I think that the best distance between the Rows is a mathematical absolute as a function of trellis height of foliage. I think the distance between the vines within the row is a function of climate, soil, vine genetics, rootstock, etc. Please note that two entirely different sets of factors are assumed by me to be basal for analysis and design. Yet, your quote directly opposes this separation of factors. Thus, my request for the assistance of your foundational thoughts.

Let me now carefully explain my thinking behind the above-noted separation. First, the "In-Row" spacing: (I am assuming vertical shoot position throughout.)

I view the fruiting wire as the "Production Line" in the "Grape Factory". To the extent that we fill it to less than its capacity we are operating below "Potential Factory Capacity", regardless of the source of that shortfall (design, vigor, etc.). The essence of this, then, is balance. Space for the individual vine must be allocated along this plane such that the vine just fills it entirely, ripens its crop, and generates its life energies for the subsequent years. The goal, in my mind, is a "continuous function" of fruit dispersal from end post to end post with no discontinuity in the pattern. Ergo, the Matrix of farming factors applies here. The genetic vigor of the varietal within the confines of the localized climate, farming techniques utilized, rootstock (if applicable – own-root characteristics if not), Soil Composition and Structure, and available water (from whatever source) are all factors affecting the vine's growth desires. It is these desires that must be controlled and directed by us – the growers.

As I am of the opinion (and have been for fifteen or more years) that excess growth is severely detrimental to wine quality, balancing of the aforementioned matrix is critical. The use of crop load to "drag–down" vigor is one tool to the farmer – a tool that has that nice characteristic of hopefully returning income to him. By distributing the fruit pattern such that a "Wall of Fruit" is formed, yet each shoot on the vine achieves 14 to 18 functioning, exposed leaves, the "Balance" is achieved.

Monterey Salinas Hwy.   Monterey, CA 93940   (408) 372-7415 Cal. 1-800-BEST VIN FAX (408) 655-1

Mr. Richard Smart
August 28, 1989
Page Two

Knife-induced uniformity is acceptable to me − to curb bull-canes and to encourage lesser canes to their fuller extension. In my mind, this condition should be achieved by two-thirds of the season. I am a strong believer in the dwarf apple tenet of "Mold and Hold". We do not wish tip extension occurring during the ripening period. Enough − this partial discussion serves to display my mind in this area.

Secondly, Row Spacing:

I wish to ignore a commonly stated concern about equipment. We can always form metal to our desires. It is a nonsensical objection at worst, a lazy one at best.

Historically, we were taught that all life on Earth was a function of photosynthesis. We now know that is not true, but functionally for our purposes it is true. The critical aspect of the previous discussion concerning "Balance" was the word "exposed", referencing leaf surface. That is, we wish the leaves exposed to sunlight, whatever the variation in design. For this discussion I will ignore basal leaf scenescence and the associated "potassium drive" − an observation on which I first spoke in 1976. It is not necessary here − affecting only design considerations.

A variety of aspects can be utilized on in-Row Trellis design considerations, but one important, but common, feature will be the height of the base wire from the ground. Whatever it is, it will determine the height of the functional foliage within the parameters of the above noted balanced vine, and will be most probably in the range of 3 to 4 inches of internode times sixteen leaves plus spur height plus starting wire level.

The basal wire height relative to the foliage height creates a simple geometry problem if we consider that the goal is no Row caused shading during useful sunlight hours. Thus, the azimuth of the sun during the forty-five days before harvest at a given site is of interest. Depending upon the chosen orientation of the Rows, directionally, a shading out zone can be calculated and the distance between the Rows determined to maximize intercepted solar radiation.

In fact, the methodology should be used in advance of layout, modified slightly by a couple of other small factors. In my opinion, land parcel configuration should not really be one of them − at least to any great degree.

More than a decade ago I did some empirical testing in the field along these lines by setting up some poles and tracking shade lines. The conclusions I came to here were thus mechanically determined and then modified by my circumstances. These observations were consistent with workers on dwarf apple design of several decades ago. Subsequently, I have been made aware of the existence of a mathematical approach to this area by workers at Geisenheim −− a study I personally have not seen. I am told that our conclusions are the same − and that is:

The distance between the Rows should be at a one to one ratio in order to maximize the potential received solar radiation. That "Ratio" means one foot distance horizontally for each one foot vertically.

Mr. Richard Smart
August 28, 1989
Page Three

The closer the rows within the aforementioned limits, the greater the yield. <u>If</u>
balance is achieved <u>down</u> <u>the</u> <u>row</u> – that is, the entire longitudinal space utilized, then
yield is purely a <u>function</u> <u>of</u> Row width not vine spacing within the Row.   The
geometric footprint wherein solar radiation reception is maximized (other factors
sufficient) is the maximum potential yield design.

I hope I have made my present thinking clear. Having done that, I'm sure you can
see my concerns with your quoted statement. It makes me wonder at my thoughts and
what you are seeing that makes us opposed along these lines. Another curious aspect
is that I interpret your previous writings as implied support of my view. A sharing of
your critical thoughts would be welcomed.

I have great respect for your work and contribution to modern Viticultural thought. I
look forward to your comments.

Sincerely,

J. Douglas Meador

JDM/cs

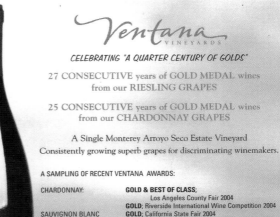

ADDENDUM
October, 2008

THE VENTANA VINEYARD
The most award winning single vineyard property in America.

- As of this writing, combining the Ventana label and the Meador Estate label, the *vineyard* has received 249 medals this *calendar year.*

- The *vineyard* is now 30 *consecutive* years of Gold medals on its Riesling grapes from and including its first harvest. No other vineyard in the world remotely has such a performance on a single variety.

- The *vineyard* is now 29 *consecutive* years of Gold medals on its Chardonnay grapes from and including its first harvest. No other vineyard in the world remotely has such a performance on two varieties.

- The *vineyard* has received Gold medals for wines from: Cabernet Sauvignon; Merlot; Cabernet Franc; Pinot Noir; Sangiovese; Tempranillo; Syrah; Grenache; Chardonnay; Sauvignon Blanc; Pinot Blanc; Riesling; Gewurztraminer; Chenin Blanc.